T0182185

FAST MULTIPOLE BOUNDARY ELEMENT METHOD

The fast multipole method is one of the most important algorithms in computing developed in the 20th century. Along with the fast multipole method, the boundary element method (BEM) has also emerged as a powerful method for modeling large-scale problems. BEM models with millions of unknowns on the boundary can now be solved on desktop computers using the fast multipole BEM. This is the first book on the fast multipole BEM, which brings together the classical theories in BEM formulations and the recent development of the fast multipole method. Two- and three-dimensional potential, elastostatic, Stokes flow, and acoustic wave problems are covered, supplemented with exercise problems and computer source codes. Applications in modeling nanocomposite materials, biomaterials, fuel cells, acoustic waves, and image-based simulations are demonstrated to show the potential of the fast multipole BEM. This book will help students, researchers, and engineers to learn the BEM and fast multipole method from a single source.

Dr. Yijun Liu has more than 25 years of research experience on the BEM for subjects including potential; elasticity; Stokes flow; and electromagnetic, elastic, and acoustic wave problems, and he has published extensively in research journals. He received his Ph.D. in theoretical and applied mechanics from the University of Illinois and, after a postdoctoral research appointment at Iowa State University, he joined the Ford Motor Company as a CAE (computer-aided engineering) analyst. He has been a faculty member in the Department of Mechanical Engineering at the University of Cincinnati since 1996. Dr. Liu is currently on the editorial board of the international journals *Engineering Analysis with Boundary Elements* and the *Electronic Journal of Boundary Elements.*

Fast Multipole
Boundary Element Method

THEORY AND APPLICATIONS
IN ENGINEERING

Yijun Liu

University of Cincinnati

CAMBRIDGE
UNIVERSITY PRESS

32 Avenue of the Americas, New York NY 10013-2473, USA

Cambridge University Press is part of the University of Cambridge.

It furthers the University's mission by disseminating knowledge in the pursuit of
education, learning and research at the highest international levels of excellence.

www.cambridge.org
Information on this title: www.cambridge.org/9781107655669

First published 2009
First paperback edition 2014

A catalogue record for this publication is available from the British Library

Library of Congress Cataloguing in Publication data
Liu, Yijun, 1959–
Fast multipole boundary element method : theory and applications in engineering /
Yijun Liu.
    p.   cm.
Includes bibliographical references and index.
ISBN 978-0-521-11659-6 (hardback)
1. Boundary element methods. I. Title.
TA347.B69L585   2009
620.001'51866 – dc22   2009010912

ISBN 978-0-521-11659-6 Hardback
ISBN 978-1-107-65566-9 Paperback

Contents

Preface

This book is an introduction to the fast multipole boundary element method (BEM), which has emerged in recent years as a powerful and practical numerical tool for solving large-scale engineering problems based on the boundary integral equation (BIE) formulations. The book integrates the classical results in BIE formulations, the conventional BEM approaches applied in solving these BIEs, and the recent fast multipole BEM approaches for solving large-scale BEM models. The topics covered in this book include potential, elasticity, Stokes flow, and acoustic wave problems in both two-dimensional (2D) and three-dimensional (3D) domains.

The book can be used as a textbook for a graduate course in engineering and by researchers in the field of applied mechanics and engineers from industries who would like to further develop or apply the fast multipole BEM to solve large-scale engineering problems in their own field. This book is based on the lecture notes developed by the author over the years for a graduate course on the BEM in the Department of Mechanical Engineering at the University of Cincinnati. Many of the results are also from the research work of the author's group at Cincinnati and from the collaborative research conducted by the author with other researchers during the last 20 years.

The book is divided into six chapters. Chapter 1 is a brief introduction to the BEM and the fast multipole method. Discussions on the advantages of the BEM are highlighted. A simple beam problem is used to illustrate the idea of transforming a problem cast in a differential equation formulation to a boundary equation formulation. The mathematical background needed in this book is also reviewed in this chapter.

Chapter 2 is on the potential problems governed by the Poisson equation or the Laplace equation. This is the most important chapter of this book, which presents the procedures in developing the BIE formulations and the conventional BEM to solve these BIEs. The fundamental solution and its properties are discussed. Both the conventional (singular) and hypersingular BIE formulations are presented, and the weakly singular nature of these BIEs is emphasized. Discretization of the BIEs using constant and higher-order elements is presented, and the related issues in handling multidomain problems, domain integrals, and indirect BIE formulations are also reviewed. Finally, programming for the conventional BEM is discussed, followed by numerical examples solved by using the conventional BEM.

Chapter 3 is on the fast multipole BEM for solving potential problems, which lays the foundations for all the subsequent chapters. Detailed derivations of the formulations, discussions on the algorithms, and computer programming for the fast multipole BEM are presented for 2D potential problems, which will serve as the prototype of the fast multipole BEM for all other problems discussed in the subsequent chapters. Then, the fast multipole formulation for 3D potential problems is presented. Numerical examples of both 2D and 3D problems are presented to demonstrate the efficiency and accuracy of the fast multipole BEM for solving large-scale problems. This chapter should be considered the focus of this book and studied thoroughly if one wishes to develop his or her own fast multipole BEM computer codes for solving other problems.

The approaches and results developed in Chapters 2 and 3 are extended in the following three chapters to solve 2D and 3D elasticity problems (Chapter 4), Stokes flow problems (Chapter 5), and acoustic wave problems (Chapter 6). In each case, the related BIE formulations are presented first, and the same systematic fast multipole BEM approaches developed for 2D and 3D potential problems are extended to the related fast multipole formulations for the subject of the chapter. In all of these chapters, the use of the dual BIE formulations (a linear combination of the conventional and hypersingular BIEs) is emphasized because of the faster convergence rate they have for the fast multipole BEM solutions.

One important objective of this book is to demonstrate the applications of the fast multipole BEM in solving large-scale practical engineering problems. To this end, many numerical examples are presented in Chapters 3–6 to demonstrate the relevance and usefulness of the fast multipole BEM, not only in academic research but also in real engineering applications. Many of the large-scale models solved by using the fast multipole BEM are still beyond the reach of the domain-based numerical methods, which clearly demonstrates the huge potentials of the fast multipole BEM in many emerging areas such as modeling of advanced composites, biomaterials, microelectromechanical systems, structural acoustics, and image-based modeling and analysis.

Exercise problems are provided at the end of each chapter for readers to review the materials covered in the chapter. More exercise problems or course projects on computer-code development and software applications can be utilized to help further understand the methods and enhance the skills. All of the computer programs of the fast multipole BEM for potential, elasticity, Stokes flow and acoustic wave problems that are discussed in this book are available from the author's website (http://urbana.mie.uc.edu/yliu).

Analytical integration of the kernel functions for 2D potential, elasticity, and Stokes flow cases and the sample computer source codes for both the 2D potential conventional BEM and the fast multipole BEM are provided in the two appendices. Electronic copies of these source codes can be downloaded from this book's webpage at the Cambridge University Press website. References for all the chapters are provided at the end of the book.

The author hopes that this book will help to advance the fast multipole BEM – an elegant numerical method that has huge potential in solving many large-scale problems in engineering. The author welcomes any comments and suggestions on further improving this book in its future editions and also takes full responsibility for any mistakes and typographical errors in this current edition.

Yijun Liu
Cincinnati, Ohio, USA
Yijun.Liu@uc.edu

Acknowledgments

The author would like to dedicate this book to Professor Frank J. Rizzo, a pioneer in the development of the BIE and BEM and now retired after teaching for more than 30 years at four universities in the United States. The author was fortunate enough to have the opportunity of conducting research under the guidance of Professor Rizzo from 1988 to 1994, first as a Ph.D. student and later as a postdoctoral research associate, at three of the four universities. His insightful views on the BIE and BEM, his serious attitude toward research, and his thoughtfulness to his students have had an immense and long-lasting impact on the author's academic career.

The author is also indebted to Professor Tianqi Ye, now retired from the Northwestern Polytechnical University in Xi'an, China, who introduced the author to the interesting subject of the BIE and BEM and taught the author that "everything important is simple" in order to pursue the best solutions for seemingly complicated problems.

The author would also like to thank Professor Naoshi Nishimura at Kyoto University for his tremendous help in the research on the fast multipole BEM in the past few years. During 2003–2004, the author spent eight months in Professor Nishimura's group and gained in-depth knowledge of the fast multipole BEM through almost daily discussions with Professor Nishimura. Much of the content presented in this book is based on the collaborative work of the author with Professor Nishimura's group at Kyoto University.

During the course of his research in the last 20 years, the author received a great deal of advice and help from many other researchers in the field of BIE and BEM. He would like to thank Professor David J. Shippy at the University of Kentucky and Professor Thomas J. Rudolphi at Iowa State University for their advice in different stages of his graduate studies, and Professor Subrata Mukherjee at Cornell University for the continued exchange of ideas and collaborations on several research endeavors that have benefited the author greatly.

The author would also like to sincerely thank his former and current students at the University of Cincinnati for their contributions to the research on the fast multipole BEM, especially to Drs. Liang Shen (3D potential and acoustics), Xiaolin Chen (image-based modeling with the fast multipole BEM), and Milind Bapat (2D and 3D acoustics). Without the students' research contributions, this book would not have been possible.

The author sincerely acknowledges the U.S. National Science Foundation for supporting his research and the Japan Society for the Promotion of Science Fellowship for Senior Researchers. Permission from Advanced CAE Research, LLC (ACR) in using the software package *FastBEM Acoustics*® for solving the 3D examples in Chapter 6 is also acknowledged.

Senior editor Peter C. Gordon at Cambridge University Press offered tremendous encouragement and advice to the author in the preparation of this manuscript. The author sincerely thanks him for his professional help in this endeavor.

Finally, the author would like to express his gratitude to his wife Rue Yuan, son Fred, and family back in China for their understanding, encouragement, patience, and sacrifice during the last 20 years.

Acronyms Used in This Book

1D:	one-dimensional
2D:	two-dimensional
3D:	three-dimensional
BC:	boundary condition
BEM:	boundary element method
BIE:	boundary integral equation
BNM:	boundary node method
CBIE:	conventional boundary integral equation
CHBIE:	dual BIE formulation
CNT:	carbon nanotube
CPU:	central processing unit
CPV:	Cauchy principal value
DOF:	degree of freedom
EFM:	element-free method
FDM:	finite difference method
FEM:	finite element method
FFT:	fast Fourier transform
FMM:	fast multipole method
GMRES:	generalized minimal residual
HBIE:	hypersingular boundary integral equation
HFP:	Hadamard finite part
L2L:	local-to-local
M2L:	moment-to-local
M2M:	moment-to-moment
M2X:	multipole-to-exponential

MD: molecular dynamics
MEMS: microelectromechanical system

NURBS: nonuniform rational B spline

ODE: ordinary differential equation

PC: personal computer
PDE: partial differential equation

Q8:   eight-node
Q4:   four-node

RAM: random-access memory
RBC: red blood cell
RVE: representative volume element

SOFC: solid oxide fuel cell
STL: stereolithography

X2L: exponential-to-local
X2X: exponential-to-exponential

1 Introduction

1.1 What Is the Boundary Element Method?

The *boundary element method* (BEM) is a numerical method for solving boundary-value or initial-value problems formulated by use of *boundary integral equations* (BIEs). In some literature, it is also called the boundary integral equation method. Figure 1.1 shows the relation of the BEM to other numerical methods commonly applied in engineering, namely the *finite difference method* (FDM), *finite element method* (FEM), *element-free* (or *meshfree*) *method* (EFM), and *boundary node method* (BNM). The FDM, FEM, and EFM can be regarded as domain-based methods that use ordinary differential equation (ODE) or partial differential equation (PDE) formulations, whereas the BEM and BNM are regarded as boundary-based methods that use the BIE formulations. It should be noted that the ODE/PDE formulation and the BIE formulation for a given problem are equivalent mathematically and represent the local and global statements of the same problem, respectively. In the BEM, only the boundaries – that is, surfaces for three-dimensional (3D) problems or curves for two-dimensional (2D) problems – of a problem domain need to be discretized. However, the BEM does have similarities to the FEM in that it does use elements, nodes, and shape functions, but on the boundaries only. This reduction in dimensions brings about many advantages for the BEM that are discussed in the following sections and throughout this book.

1.2 Why the Boundary Element Method?

The BEM offers some unique advantages for solving many engineering problems. The following are the main advantages of the BEM:

- *Accuracy:* The BEM is a semianalytical method and thus is more accurate, especially for stress concentration problems such as fracture analysis of structures.

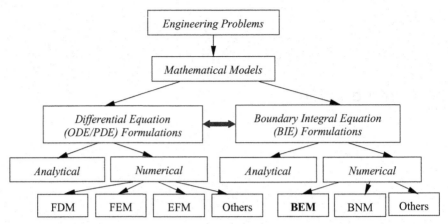

Figure 1.1. Relations of commonly used numerical methods for solving engineering problems.

- *Efficient in modeling:* The BEM mesh (a collection of the elements used to discretize a continuum structure) is much easier to generate for 3D problems or infinite domain problems because of the dimension reduction in the BIE formulations.
- *An independent numerical method:* The BEM can be applied along with the other domain-based methods to verify the solutions to a problem for which no analytical solution is available.

1.3 A Comparison of the Finite Element Method and the Boundary Element Method

Table 1.1 gives a comparison of the BEM with the FEM regarding their main features, as well as advantages and disadvantages. This comparison is by no

Table 1.1. *A comparison of the FEM and BEM*

FEM	BEM
Features	
• Derivative-based (local) approach	• Integral-based (global) approach
• Domain mesh: 2D or 3D mesh	• Boundary mesh: 1D or 2D mesh
• Symmetrical, sparse matrices	• Nonsymmetrical, dense matrices
• Many commercial packages available	• Fewer commercial packages available
Advantages	
• Solution is fast	• Mesh generation is fast
• Suitable for general structure analysis; large mechanical systems	• Suitable for stress concentration problems (e.g., fracture mechanics)
• Nonlinear problems	• Infinite domain problems
• Composite materials (macroscale analysis)	• Composite materials (e.g., microscale continuum models)

means complete, and certainly will change with the new development in either the FEM or BEM.

1.4 A Brief History of the Boundary Element Method and Other References

The *direct* BIE formulations and their modern numerical solutions that use boundary elements for problems in applied mechanics originated more than 40 years ago during the 1960s. The 2D potential problem was first formulated in terms of a direct BIE and solved numerically by Jaswon [1], Symm [2], and Jaswon and Ponter [3]. This work was later extended to the vector case – 2D elastostatic problem by Rizzo in the early 1960s for his Ph.D. dissertation at the University of Illinois at Urbana-Champaign, which was later published as a journal article in 1967 [4]. Following these early works, extensive research efforts were made in BIE formulations of many problems in applied mechanics and in the numerical solutions during the 1960s and 1970s [5–20]. The name *boundary element method* appeared in the mid-1970s in an attempt to make an analogy with the FEM [21–23].

Some of the important textbooks and research volumes in the 1980s and early 1990s, which made significant contributions to the research and development of the BIE/BEM, can be found in Refs. [24–28]. A few recent research volumes with advanced treatment of the topics on BIE/BEM can be found in Refs. [29–32]. Readers may consult these publications for more detailed discussions on many of the topics in this book or other topics not covered in this book regarding the BIE formulations and the related conventional BEM solution techniques.

1.5 Fast Multipole Method

Although the BEM has enjoyed the reputation of easy meshing in modeling many problems with complicated geometries, its efficiency in solutions has been a serious problem for analyzing large-scale models. For example, the BEM has been limited to solving problems with a few thousand degrees of freedom (DOFs) on a personal computer (PC) for many years. This is because the conventional BEM, in general, produces dense and nonsymmetric matrices that, although smaller in size, require $O(N^2)$ operations to compute the coefficients and another $O(N^3)$ operations to solve the system by using direct solvers (here, N is the number of equations of the linear system or DOFs in the BEM model).

In the mid-1980s, Rokhlin and Greengard [33–35] pioneered the innovative *fast multipole method* (FMM) that can be used to accelerate the solutions of BIE by severalfold to reduce the CPU time in a FMM-accelerated BEM to $O(N)$. However, it took almost a decade for the mechanics community to realize the potential of the FMM for the BEM. Some of the early research on the fast multipole BEM in applied mechanics can be found in Refs. [36–40], which show the great promise of the fast multipole BEM for solving large-scale engineering problems. A comprehensive review of the fast-multipole-accelerated BIE/BEM and the research work up to 2002 can be found in Ref. [41].

In this book, we use the FMM to solve the various BEM systems of equations for potential, elastostatic, Stokes flow, and acoustic wave problems. The fast multipole BEM represents the future of BEM research and applications. However, understanding the BIE formulations and the conventional BEM procedures in solving these BIEs is still very important. Learning the intricacies of the BIE formulations and the conventional BEM while promoting the fast multipole BEM is emphasized in this book.

1.6 Applications of the Boundary Element Method in Engineering

Today, the BEM has gained a great deal of attention in the field of computational mechanics, especially with the help of the FMM. The applications of the BEM are now well beyond the range of classical potential and elasticity theories, extending to many engineering fields, including heat transfer, diffusion and convection, fluid flows, fracture mechanics, geomechanics, plates and shells, inelastic problems, contact problems, wave propagations (acoustic, elastic, and electromagnetic waves), electrostatic problems, design sensitivity and optimizations, and inverse problems. Examples of the fast multipole BEM applications are given in the following chapters, in which applications of the fast multipole BEM for solving large-scale problems in many engineering fields are presented.

As an example, we use an engine-block model (Figure 1.2) to conduct a thermal analysis and compare the results obtained with the FEM and the BEM. With the FEM (using ANSYS®), more than 363,000 volume elements are applied with DOFs above 1.5 million. With the BEM (a fast multipole BEM code discussed in Chapter 3), only about 42,000 constant surface elements (triangular constant elements) are applied with the same number of DOFs. Furthermore, meshing the volume is considerably more difficult and takes longer human time than meshing the surfaces of the engine block. On a desktop PC, the FEM solution took 50 min to finish, whereas the BEM solution took only about 16 min. The differences in the computed results for the temperature fields by the FEM and the BEM (Figure 1.3) are less than 1%. Considering the human time saved during the discretization stage, the advantage of the BEM in modeling 3D problems with complicated geometries is most evident.

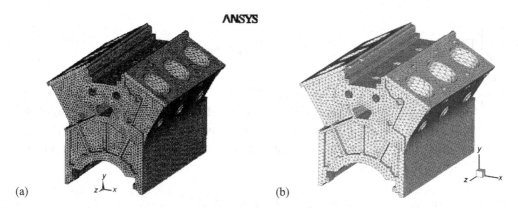

Figure 1.2. An engine block discretized using finite elements and boundary elements: (a) FEM (363,000 volume elements/1.5 million DOFs), (b) BEM (42,000 surface elements/DOFs).

1.7 An Example – Bending of a Beam

We first study a simple beam-bending problem (Figure 1.4) to see that the boundary approach is a valid and equivalent approach to solving engineering problems that are usually written in ODEs or PDEs.

We have the following governing equations based on simple beam theory:

$$EI\frac{d^2v}{dx^2} = M(x),\tag{1.1}$$

$$\frac{dM}{dx} = Q(x),\tag{1.2}$$

$$\frac{dQ}{dx} = q(x),\tag{1.3}$$

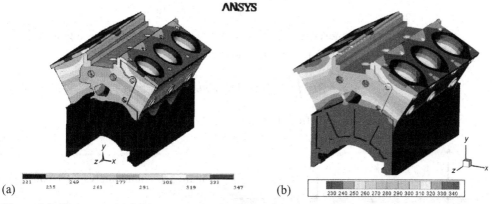

Figure 1.3. Temperature field computed using finite elements and boundary elements: (a) FEM (CPU time = 50 min), (b) BEM (CPU time = 16 min).

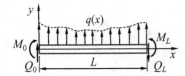

Figure 1.4. A simple beam-bending problem.

for $x \in (0, L)$, where $v(x)$ is the deflection of the beam, EI is the bending stiffness, $M(x)$ is the bending moment, $Q(x)$ is the shear force, and $q(x)$ is the distributed load in the lateral direction (Figure 1.4). Combining Eqs. (1.1)–(1.3), we also have:

$$EI\frac{d^4v}{dx^4} = q(x). \tag{1.4}$$

To solve the beam problem, we need to solve either Eq. (1.1) if the bending moment $M(x)$ is known or Eq. (1.4) if $M(x)$ is not readily available, under given boundary conditions at $x = 0$ and $x = L$. In the following discussion, it is shown that solving ODE (1.1) is equivalent to solving an integral equation formulation that involves boundary values only.

We first consider the so-called *fundamental solution* for Eq. (1.1), or *the Green's function* for an infinitely long beam (Figure 1.5). Consider the load case in which a unit concentrated force $P = 1$ is applied at point x_0 of the beam.

The bending moment $M^*(x_0, x)$ in the beam at x is governed by the following equation [see Eqs. (1.2) and (1.3)]:

$$\frac{d^2 M^*(x_0, x)}{dx^2} = \delta(x_0, x), \quad \forall x, x_0 \in (-\infty, +\infty), \tag{1.5}$$

where $\delta(x_0, x)$ is the Dirac δ function used to represent the distributed load $q(x)$ in this case. An engineering "definition" of the Dirac δ function $\delta(x_0, x)$ can be given as:

$$\delta(x_0, x) = \begin{cases} 0, & \text{if } x \neq x_0 \\ \infty, & \text{if } x = x_0 \end{cases}. \tag{1.6}$$

An important property of the Dirac δ function $\delta(x_0, x)$, which is a generalized function, is the sifting property [42] given by:

$$\int_{-\infty}^{+\infty} f(x)\delta(x_0, x)dx = f(x_0) \tag{1.7}$$

for any continuous function $f(x)$.

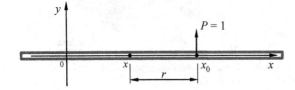

Figure 1.5. An infinitely long beam with a point force.

Solving Eq. (1.5) by using, for example, Fourier transformation (see Problem 1.1) or simply from the physical argument, we can show that the bending moment at x that is due to the unit point force at x_0 is:

$$M^*(x_0, x) = \frac{1}{2}r, \tag{1.8}$$

where $r = |x_0 - x|$ is the distance between the *source point* x_0 and *field point* x. This is the fundamental solution for Eq. (1.1) and is the first ingredient needed in our boundary formulation. The second ingredient is the following generalized Green's identity:

$$\int_0^L \left(u \frac{d^2v}{dx^2} - \frac{d^2u}{dx^2} v \right) dx = \left(u \frac{dv}{dx} - \frac{du}{dx} v \right) \Big|_{x=0}^{x=L} \tag{1.9}$$

for any two functions $u(x)$ and $v(x)$ with sufficient smoothness (continuity of the derivatives). The significance of this identity is that it can transform a one-dimensional (1D) domain integral to evaluations of the functions at the boundaries.

Now if we select u to be the fundamental solution $M^*(x_0, x)$ satisfying Eq. (1.5) and v to be the deflection of the beam satisfying Eq. (1.1), we have the following result from Eq. (1.9):

$$\int_0^L \left(M^* \frac{d^2v}{dx^2} - \frac{d^2 M^*}{dx^2} v \right) dx = \left(M^* \frac{dv}{dx} - \frac{d M^*}{dx} v \right) \Big|_{x=0}^{x=L}.$$

Applying Eqs. (1.1) and (1.5), we obtain

$$v(x_0) = \int_0^L \left(M^* \frac{M}{EI} \right) dx - \left(M^* \frac{dv}{dx} - \frac{d M^*}{dx} v \right) \Big|_{x=0}^{x=L}$$

or

$$v(x_0) = \int_0^L M^*(x_0, x) \frac{M(x)}{EI} dx + Q^*(x_0, L)v_L - Q^*(x_0, 0)v_0$$
$$- M^*(x_0, L)\theta_L + M^*(x_0, 0)\theta_0, \quad \forall x_0 \in (0, L), \tag{1.10}$$

in which v_0, v_L, θ_0, and θ_L are the deflection and rotation of the beam at the left and right ends, respectively, and Q^* is the shear force in the fundamental solution corresponding to M^* in (1.8); that is:

$$Q^*(x_0, x) = \frac{d M^*(x_0, x)}{dx} = \begin{cases} \dfrac{1}{2}, & \text{for } x > x_0 \\[2mm] -\dfrac{1}{2}, & \text{for } x < x_0 \end{cases}. \tag{1.11}$$

Equation (1.10) is an expression of the solution for deflection at any point inside the beam. Once the deflections and rotations at the two ends (boundaries) of the beam are obtained, we can use Eq. (1.10) to evaluate the deflection of the beam at any point x_0.

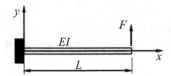

Figure 1.6. A cantilever beam.

To derive a boundary formulation, we first let x_0 tend to 0 in Eq. (1.10) to have:

$$v_0 = \int_0^L \frac{x}{2} \frac{M(x)}{EI} dx + \frac{1}{2} v_L + \frac{1}{2} v_0 - \frac{L}{2} \theta_L,$$

and then we let x_0 tend to L in Eq. (1.10) to have:

$$v_L = \int_0^L \frac{L-x}{2} \frac{M(x)}{EI} dx + \frac{1}{2} v_L + \frac{1}{2} v_0 + \frac{L}{2} \theta_0.$$

Writing the two equations in a matrix form, we obtain the following *boundary formulation*:

$$\frac{1}{2} \begin{bmatrix} 1 & -1 \\ -1 & 1 \end{bmatrix} \begin{Bmatrix} v_0 \\ v_L \end{Bmatrix} + \frac{L}{2} \begin{bmatrix} 0 & 1 \\ -1 & 0 \end{bmatrix} \begin{Bmatrix} \theta_0 \\ \theta_L \end{Bmatrix} = \frac{1}{2EI} \int_0^L \begin{Bmatrix} x \\ L-x \end{Bmatrix} M(x) dx.$$

$$(1.12)$$

This boundary formulation is equivalent to the ODE given in (1.1). If the bending moment is known, this equation can be applied to solve for the unknown boundary variables v_0, v_L, θ_0, and θ_L first.

As an example, we consider the cantilever beam in Figure 1.6 by using our derived boundary formulation. In this case, the bending moment is found to be:

$$M(x) = F(L-x),$$

and the boundary conditions are:

$$v_0 = 0, \quad \theta_0 = 0.$$

Thus, boundary equation (1.12) yields:

$$\frac{1}{2} \begin{bmatrix} -1 & L \\ 1 & 0 \end{bmatrix} \begin{Bmatrix} v_L \\ \theta_L \end{Bmatrix} = \frac{FL^3}{12EI} \begin{Bmatrix} 1 \\ 2 \end{Bmatrix}.$$

Solving this equation, we obtain the deflection and rotation of the beam at the right end:

$$\begin{Bmatrix} v_L \\ \theta_L \end{Bmatrix} = \frac{FL^3}{6EI} \begin{Bmatrix} 2 \\ 3/L \end{Bmatrix}.$$

Substituting these results into expression (1.10), we also have:

$$v(x_0) = \int_0^L \frac{|x - x_0|}{2} \frac{F(L - x)}{EI} dx + \frac{1}{2} \left(\frac{FL^3}{3EI} \right) - \frac{L - x_0}{2} \left(\frac{FL^2}{2EI} \right)$$

$$= \frac{F}{6EI}(3L - x_0)x_0^2, \quad \forall x_0 \in (0, L);$$

which agrees with the result from solving Eq. (1.1) directly. Thus, boundary formulation (1.12) is equivalent to the ODE formulation in Eq. (1.1).

Note that the simple beam example is used here to illustrate the procedures in transforming an ODE or PDE statement of a problem to a boundary formulation and the ingredients needed in this process. It does not mean that we will use this boundary formulation to solve beam-bending problems. In fact, there are no advantages in solving 1D problems by using the boundary formulations or the BEM in general.

The two major ingredients in the boundary formulation are the fundamental solution and the generalized Green's identity. These two topics are expanded in following sections.

1.8 Some Mathematical Preliminaries

Some mathematical results needed in later chapters of this book are reviewed in this section. For more detailed coverage of these topics, the reader should consult other books on the related topics. Many of the topics are covered in Fung's outstanding textbook [43].

1.8.1 Integral Equations

An integral equation is an equation that contains unknown functions under the integral sign. For example, the following equations are two integral equations in one dimension:

$$\int_a^b K(x, y)\phi(y)dy = f(x), \tag{1.13}$$

$$\phi(x) = \int_a^b K(x, y)\phi(y)dy + g(x), \tag{1.14}$$

in which ϕ is an unknown function, $K(x, y)$ is a known *kernel* function, and f and g are two given functions. Equation (1.13) is a linear *Fredholm equation of the first kind*, whereas Eq. (1.14) is a linear *Fredholm equation of the second kind*. The kernel function $K(x, y)$ determines the characteristics of the integral equation. For example, if:

$$K(x, y) = \frac{1}{|x - y|},$$

then the integrals in (1.13) and (1.14) are singular when $x \in (a, b)$, and Eqs. (1.13) and (1.14) are called singular integral equations.

1.8.2 Indicial Notation

Indicial notation is extremely useful in deriving the equations in BIE formulations. In indicial notation, coordinates x, y, and z are replaced with x_1, x_2, and x_3, respectively, for 3D problems, or simply as x_i, for $i = 1, 2$ (for two dimensions) or $1, 2, 3$ (for three dimensions). For example, the equation of a plane in 3D space, $ax + by + cz = p$, can be written as:

$$\sum_{i=1}^{3} a_i x_i = p,$$

if we set $a_1 = a$, $a_2 = b$, and $a_3 = c$. The preceding expression can be further simplified if we apply *Einstein's summation convention*, which says that summation is implied if an index is repeated twice in the same term. With this convention, the preceding equation for the plane in 3D space can be written simply as:

$$a_i x_i = p,$$

where i is called a dummy index and can be changed to other symbols. For example, the dot product of two vectors $\vec{a}$ and $\vec{b}$ can be expressed as:

$$\vec{a} \cdot \vec{b} = a_i b_i = a_k b_k,$$

in indicial notation. A linear system of equations $\mathbf{Ax} = \mathbf{b}$ can be written as:

$$a_{ij} x_j = b_i,$$

with indices i and j running from $1, 2, \ldots, n$ (number of the equations).

Differentiations of a function $f(x, y, z) = f(x_i)$ can be expressed as:

$$\frac{\partial f}{\partial x}, \frac{\partial f}{\partial y}, \frac{\partial f}{\partial z} \quad \Rightarrow \quad \frac{\partial f}{\partial x_i} \equiv f_{,i},$$

$$df = \frac{\partial f}{\partial x_1} dx_1 + \frac{\partial f}{\partial x_2} dx_2 + \frac{\partial f}{\partial x_3} dx_3 = f_{,i} \, dx_i,$$

$$\nabla^2 f = \frac{\partial^2 f}{\partial x_1^2} + \frac{\partial^2 f}{\partial x_2^2} + \frac{\partial^2 f}{\partial x_3^2} = f_{,ii}. \tag{1.15}$$

The *Kronecker delta* δ_{ij} is defined by:

$$\delta_{ij} = \begin{cases} 1, & \text{if } i = j \\ 0, & \text{if } i \neq j \end{cases}, \tag{1.16}$$

which is similar to the identity matrix. The Kronecker delta can be used to simplify expressions. For example,

$$a_i b_j \delta_{ij} = a_i b_i = a_j b_j \text{ and } f_{,ij}\,\delta_{jk} = f_{,ik}.$$

Another important symbol in indicial notation is the *permutation symbol* e_{ijk}, which is defined as:

$$e_{ijk} = \begin{cases} 1, & \text{for cyclic suffix order: } 123, 231, 312 \\ -1, & \text{for cyclic suffix order: } 132, 213, 321 \\ 0, & \text{if any two indices are the same.} \end{cases} \tag{1.17}$$

For example, $e_{112} = 0$, $e_{231} = 1$, $e_{213} = -1$, $e_{333} = 0$, and so on. The vector product of two vectors $\vec{a}$ and $\vec{b}$ is $\vec{c} = \vec{a} \times \vec{b}$. In indicial notation, the components of $\vec{c}$ are given by $c_i = e_{ijk}a_j b_k$ when the permutation symbol is used.

A useful relation between the Kronecker delta and the permutation symbol is:

$$e_{ijk}e_{ilm} = \delta_{jl}\delta_{km} - \delta_{jm}\delta_{kl}. \tag{1.18}$$

This relation can be verified from the vector identity:

$$\vec{a} \times (\vec{b} \times \vec{c}) = (\vec{a} \cdot \vec{c})\vec{b} - (\vec{a} \cdot \vec{b})\vec{c}.$$

1.8.3 Gauss Theorem

The Gauss theorem in calculus is probably the single most important formula we need in the development of BIE formulations. For a closed domain V (either in two or three dimensions) with boundary S, we have:

$$\int_V \phi_{,i}\, dV = \int_S \phi n_i\, dS \tag{1.19}$$

for any differentiable function $\phi(x_i)$, where n_i is the component (direction cosines) of the outward normal. The following equations are some of the variations of the Gauss theorem:

$$\int_V F_{i,j}\, dV = \int_S F_i n_j\, dS, \tag{1.20}$$

$$\int_V \text{div}\, \mathbf{F}\, dV = \int_S \mathbf{F} \cdot \mathbf{n}\, dS, \tag{1.21}$$

$$\int_V \nabla \times \mathbf{F}\, dV = \int_S \mathbf{n} \times \mathbf{F}\, dS, \tag{1.22}$$

where $\mathbf{F} = F_i(x_j)$ is a vector function.

1.8.4 The Green's Identities

Using the Gauss theorem, we can establish readily the following *Green's first identity*:

$$\int_V u\nabla^2 v dV = \int_S u\frac{\partial v}{\partial n}dS - \int_V u_{,i}\, v_{,i}\, dV, \tag{1.23}$$

and *the Green's second identity*:

$$\int_V (u\nabla^2 v - v\nabla^2 u)dV = \int_S \left(u\frac{\partial v}{\partial n} - v\frac{\partial u}{\partial n}\right)dS \tag{1.24}$$

for any two continuous functions u and v. Various forms of the Green's second identity are used in the development of the BIEs for different problems.

1.8.5 Dirac δ Function

The Dirac δ function $\delta(\mathbf{x}, \mathbf{y})$ in two and three dimensions has the following sifting properties [42]:

$$\int_V f(\mathbf{y})\delta(\mathbf{x}, \mathbf{y})dV(\mathbf{y}) = \begin{cases} f(\mathbf{x}), & \text{if } \mathbf{x} \in V \\ 0, & \text{if } \mathbf{x} \notin V \cup S \end{cases}, \tag{1.25}$$

$$\int_V f(\mathbf{y})\frac{\partial}{\partial x_i}\delta(\mathbf{x}, \mathbf{y})dV(\mathbf{y}) = \begin{cases} -\dfrac{\partial}{\partial x_i} f(\mathbf{x}), & \text{if } \mathbf{x} \in V \\ 0, & \text{if } \mathbf{x} \notin V \cup S, \end{cases} \tag{1.26}$$

in which $\mathbf{x}$ and $\mathbf{y}$ represent two points in space, and $f(\mathbf{x}) = f(x_i)$ is a differentiable function. In generalized function theory, the Dirac δ function is continuous and differentiable [42]. Applications of the Dirac δ function can greatly simplify the derivations of the BIEs.

1.8.6 Fundamental Solutions

Fundamental solutions are important ingredients in BIE formulations. Without these fundamental solutions, we cannot convert the ODEs or PDEs into BIEs in general. For different problems, we have different fundamental solutions, which are the solutions that are due to a unit source (heat source, point force, unit charge, and so on) in an infinite space. These solutions have been found for most linear problems, and we do not delve into the derivations of these fundamental solutions. However, understanding the behaviors of the fundamental solution for a particular problem at hand is very important in developing good strategy to solving the problem with the BEM. This point is elaborated on in later chapters.

For simple problems, a Fourier transform can be applied to obtain the fundamental solutions. For example, for beam equation (1.4), the fundamental solution $v^*(x_0, x)$ satisfies the following equation:

$$EI\frac{d^4v^*(x_0, x)}{dx^4} = \delta(x_0, x), \quad \forall x, x_0 \in (-\infty, +\infty), \tag{1.27}$$

in which the Dirac δ function $\delta(x_0, x)$ represents the unit point force at x_0 (Figure 1.5). For a function $f(x)$, the Fourier transform and its inverse are defined by:

$$\mathcal{F}[f(x)] = \frac{1}{\sqrt{2\pi}} \int_{-\infty}^{+\infty} f(x)e^{i\lambda x}dx = \mathcal{F}(\lambda), \tag{1.28}$$

$$f(x) = \frac{1}{\sqrt{2\pi}} \int_{-\infty}^{+\infty} \mathcal{F}(\lambda) e^{-i\lambda x}d\lambda, \tag{1.29}$$

respectively. Applying the Fourier transform to Eq. (1.27) and noticing that

$$\mathcal{F}[\delta(x)] = 1, \quad \mathcal{F}\left[\frac{dv^*}{dx}\right] = i\lambda\mathcal{F}[v^*],$$

we obtain from Eq. (1.27):

$$\lambda^4 EI\mathcal{F}[v^*] = 1 \quad \text{or} \quad \mathcal{F}[v^*] = \frac{1}{EI\lambda^4}.$$

The inverse transform yields:

$$v^*(x_0, x) = \frac{1}{12EI}r^3, \quad \text{with} \quad r = |x_0 - x|. \tag{1.30}$$

This is the deflection of the beam at x that is due to the point force at x_0. Applying Eq. (1.1), we have:

$$M^*(x_0, x) = EI\frac{d^2v^*}{dx^2} = \frac{1}{2}r,$$

which is the corresponding moment in the fundamental solution as given in Eq. (1.8).

1.8.7 Singular Integrals

We encounter various so-called singular integrals in the BIE formulations. In these singular integrals, the integrands have singular points at which the integrands tend to infinity. Although we can show in later chapters that singular integrals in the BIEs can be removed analytically by use of the so-called weakly singular forms of the BIEs, understanding the singular integrals is still very important in studying BIEs and BEMs.

We use a few 1D cases as examples to illustrate the behaviors and results of the singular integrals. First, consider the following integral:

$$f_1(x) = \int_a^b \log|x - y|dy \quad \text{for } a < x < b. \tag{1.31}$$

The integrand tends to infinity at $x = y$; thus, the integral is singular. This is an *improper integral* and is evaluated as follows:

$$f_1(x) = \lim_{\varepsilon_1 \to 0} \int_a^{x-\varepsilon_1} \log|x - y| \, dy + \lim_{\varepsilon_2 \to 0} \int_{x+\varepsilon_2}^b \log|x - y| \, dy$$

$$= \lim_{\varepsilon_1 \to 0} \left[-(x - y) \log(x - y) - y \right]_{y=a}^{y=x-\varepsilon_1}$$

$$+ \lim_{\varepsilon_2 \to 0} \left[(y - x) \log(y - x) - y \right]_{y=x+\varepsilon_2}^{y=b}$$

$$= (x - a) \left[\log(x - a) - 1 \right] + (b - x) \left[\log(b - x) - 1 \right].$$

Thus, integral $f_1(x)$ in (1.31) exists regardless of the values of ε_1 and ε_2 and is called a *weakly singular* integral.

Next, consider the following strongly singular integral:

$$f_2(x) = \int_a^b \frac{1}{y - x} \, dy \quad \text{for } a < x < b. \tag{1.32}$$

We regard this as an improper integral and evaluate it as follows:

$$f_2(x) = \lim_{\varepsilon_1 \to 0} \int_a^{x-\varepsilon_1} \frac{1}{y - x} \, dy + \lim_{\varepsilon_2 \to 0} \int_{x+\varepsilon_2}^b \frac{1}{y - x} \, dy$$

$$= \lim_{\varepsilon_1 \to 0} \left[\log|y - x| \right]_{y=a}^{y=x-\varepsilon_1} + \lim_{\varepsilon_2 \to 0} \left[\log|y - x| \right]_{y=x+\varepsilon_2}^{y=b}$$

$$= \log\left(\frac{b - x}{x - a} \right) + \lim_{\varepsilon_1 \to 0} \log \varepsilon_1 - \lim_{\varepsilon_2 \to 0} \log \varepsilon_2,$$

which does not exist if ε_1 and ε_2 are kept independent of each other. It is only when $\varepsilon_1 = \varepsilon_2$ that the integral has a finite value:

$$f_2(x) = \log\left(\frac{b - x}{x - a} \right), \tag{1.33}$$

which is called the *Cauchy principal value* (CPV) of the integral in (1.32). Therefore, $f_2(x)$ is called a CPV integral in that the integral is evaluated with a small "symmetrical" region subtracted from the domain of integration $(\varepsilon_1 = \varepsilon_2)$ at x.

Consider the following hypersingular singular integral:

$$f_3(x) = \int_a^b \frac{1}{(y - x)^2} \, dy, \quad \text{for } a < x < b. \tag{1.34}$$

We evaluate this integral by using the CPV definition:

$$f_3(x) = \lim_{\varepsilon \to 0} \left[\int_a^{x-\varepsilon} \frac{1}{(y - x)^2} \, dy + \int_{x+\varepsilon}^b \frac{1}{(y - x)^2} \, dy \right]$$

$$= \lim_{\varepsilon \to 0} \left[-\frac{1}{(y - x)} \Big|_a^{x-\varepsilon} - \frac{1}{(y - x)} \Big|_{x+\varepsilon}^b \right]$$

$$= -\frac{1}{x - a} - \frac{1}{b - x} + \lim_{\varepsilon \to 0} \left(\frac{2}{\varepsilon} \right),$$

which does not exist even in the sense of a CPV integral. However, in the BIE formulations, we find that an infinite term like $2/\varepsilon$ is canceled out by the integral with the same integrand on the small region $(x - \varepsilon, x + \varepsilon)$. Therefore, $f_3(x)$ is still meaningful and called a *Hadamard finite part* (HFP) integral [44], with the finite part given by [45]:

$$f_3(x) = -\frac{1}{x-a} - \frac{1}{b-x}. \tag{1.35}$$

1.9 Summary

In this chapter, a general introduction of the BIE and the BEM is provided. A comparison of the BEM with the FEM is discussed. A simple beam problem is used as an example to show the procedures in formulating and solving a problem by using the boundary formulation. Two important ingredients are needed in the BIE formulations. One is the fundamental solution that is specific to a given problem and is available for most linear problems. Another ingredient is the generalized Green's identity associated with the differential operator for describing the problem. Some mathematical results that are needed in the development of the BIE and the BEM are reviewed, especially the index notation and the Gauss theorem in various forms.

Problems

1.1. Using a Fourier transform, solve Eq. (1.5) to obtain the moment in the fundamental solution given in (1.8) for the simple beam problem.

1.2. Derive the generalized Green's identity given in (1.9).

1.3. Derive the following generalized Green's identity corresponding to ODE (1.4),

$$\int_0^L \left[u\frac{d^4v}{dx^4} - \frac{d^4u}{dx^4}v \right] dx = \left[u\frac{d^3v}{dx^3} - \frac{d^3u}{dx^3}v + \frac{d^2u}{dx^2}\frac{dv}{dx} - \frac{du}{dx}\frac{d^2v}{dx^2} \right] \Big|_{x=0}^{x=L}, \tag{1.36}$$

for any two *continuous* functions u and v on the interval $(0, L)$. If u and v represent the deflections of a straight beam with length L, bending stiffness EI, and under two different sets of loading conditions, respectively, what is the physical meaning of this identity?

1.4. Give the values of the following expressions, if defined:

$$\delta_{ij} = ?; \quad \delta_{ij}\delta_{ij} = ?; \quad \delta_{ij}\delta_{ij}\delta_{ij} = ?.$$

1.5. Verify the following results:

$$e_{ijk}e_{ijk} = 6; \quad e_{ijk}A_jA_k = 0.$$

1.6. Express the triple scalar product $\vec{u} \cdot (\vec{v} \times \vec{w})$ of three vectors $\vec{u}$, $\vec{v}$, and $\vec{w}$ (in three dimensions) in the index form.

1.7. Verify Eq. (1.18) using the vector identity $\vec{a} \times (\vec{b} \times \vec{c}) = (\vec{a} \cdot \vec{c})\vec{b} - (\vec{a} \cdot \vec{b})\vec{c}$.

1.8. Write Eqs. (1.21) and (1.22) in index forms.

1.9. Show that the CPV of the following integral does not exist:

$$f(x) = \int_a^b \frac{1}{|y - x|} dy, \quad \text{for } a < x < b.$$

2 Conventional Boundary Element Method for Potential Problems

Many problems in engineering can be described by the Laplace equation or the Poisson equation. These problems can be termed potential problems, such as heat conduction, potential flows, electrostatic fields, or the mechanics problem of a bar in torsion. In this chapter, we study the BIE formulations for solving potential problems and learn how to solve these BIEs by using the conventional BEM. In Chapter 3, we study the fast multipole BEM that can accelerate the BEM solutions for large-scale potential problems.

2.1 The Boundary-Value Problem

We consider the following Poisson equation governing the potential field ϕ in domain V (either 2D or 3D, finite or infinite):

$$\nabla^2 \phi + f = 0, \quad \text{in } V, \tag{2.1}$$

where f is a known function in domain V. The boundary conditions (BCs) to be considered are:

$$\phi = \overline{\phi}, \quad \text{on } S_\phi \quad \text{(Dirichlet BC)}, \tag{2.2}$$

$$q \equiv \frac{\partial \phi}{\partial n} = \overline{q}, \quad \text{on } S_q \quad \text{(Neumann BC)}, \tag{2.3}$$

in which the over bar indicates the prescribed value for the function, $S_\phi \cup S_q = S$ is the boundary of the domain, and n is the outward normal of the boundary S (Figure 2.1). Note that the normal derivative of ϕ (corresponding to heat flux in thermal analysis) can be expressed as $q = \frac{\partial \phi}{\partial x_k} n_k = \phi_{,k} n_k$ in index notation, with n_k being the components or direction cosines of normal n.

With the fundamental solution and the second Green's identity, we can convert the preceding boundary-value problem given in Eqs. (2.1)–(2.3) into BIE formulations.

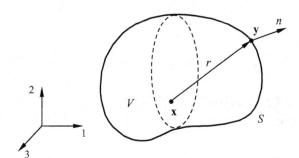

Figure 2.1. A 3D finite domain V with boundary S.

2.2 Fundamental Solution for Potential Problems

The fundamental solution $G(\mathbf{x}, \mathbf{y})$ for potential problems satisfies:

$$\nabla^2 G(\mathbf{x}, \mathbf{y}) + \delta(\mathbf{x}, \mathbf{y}) = 0, \quad \forall \mathbf{x}, \mathbf{y} \in R^2/R^3, \tag{2.4}$$

in which the derivatives are taken at point $\mathbf{y}$, that is, $\nabla^2 = \partial^2 (\cdot) / \partial y_i \partial y_i$, and R^2 and R^3 indicate the full 2D and 3D spaces, respectively. The Dirac δ function $\delta(\mathbf{x}, \mathbf{y})$ in Eq. (2.4) represents a unit source (e.g., heat source) at the *source point* $\mathbf{x}$, and $G(\mathbf{x}, \mathbf{y})$ represents the response (e.g., temperature) at the *field point* $\mathbf{y}$ that is due to that source.

The fundamental solution $G(\mathbf{x}, \mathbf{y})$ is given by:

$$G(\mathbf{x}, \mathbf{y}) = \begin{cases} \dfrac{1}{2\pi} \log\left(\dfrac{1}{r}\right), & \text{for two dimensions,} \\[2ex] \dfrac{1}{4\pi r}, & \text{for three dimensions,} \end{cases} \tag{2.5}$$

where r is the distance between the source point $\mathbf{x}$ and field point $\mathbf{y}$, and its normal derivative is:

$$F(\mathbf{x}, \mathbf{y}) \equiv \frac{\partial G(\mathbf{x}, \mathbf{y})}{\partial n(\mathbf{y})} = \begin{cases} -\dfrac{1}{2\pi r} r_{,k}\, n_k(\mathbf{y}), & \text{for two dimensions,} \\[2ex] -\dfrac{1}{4\pi r^2} r_{,k}\, n_k(\mathbf{y}), & \text{for three dimensions,} \end{cases} \tag{2.6}$$

with $r_{,k} = \partial r / \partial y_k = (y_k - x_k)/r$. The fundamental solution satisfies the following *integral identities* [46–48]:

First identity:

$$\int_S F(\mathbf{x}, \mathbf{y})\, dS(\mathbf{y}) = \begin{cases} -1, & \forall \mathbf{x} \in V \\[1ex] 0, & \forall \mathbf{x} \in E. \end{cases} \tag{2.7}$$

Second identity:

$$\int_S \frac{\partial F(\mathbf{x}, \mathbf{y})}{\partial n(\mathbf{x})}\, dS(\mathbf{y}) = 0, \quad \forall \mathbf{x} \in V \cup E. \tag{2.8}$$

Third identity:

$$\int_S \frac{\partial G(\mathbf{x}, \mathbf{y})}{\partial n(\mathbf{x})} n_k(\mathbf{y}) dS(\mathbf{y}) - \int_S \frac{\partial F(\mathbf{x}, \mathbf{y})}{\partial n(\mathbf{x})} (y_k - x_k) dS(\mathbf{y}) = \begin{cases} n_k(\mathbf{x}), & \forall \mathbf{x} \in V \\ 0, & \forall \mathbf{x} \in E. \end{cases}$$

(2.9)

Fourth identity:

$$\int_S F(\mathbf{x}, \mathbf{y})(y_k - x_k) \, dS(\mathbf{y}) - \int_S G(\mathbf{x}, \mathbf{y}) n_k(\mathbf{y}) dS(\mathbf{y}) = 0, \quad \forall \mathbf{x} \in V \cup E,$$

(2.10)

in which S can be an arbitrary *closed* contour (for two dimensions) or surface (for three dimensions), V is the domain enclosed by S, and E is the infinite domain outside S. These identities have clear physical meanings and can be very convenient in deriving various weakly singular or nonsingular forms of the BIEs for potential problems [46–48]. We can obtain these identities readily by integrating governing equation (2.4) over the domain V and invoking the Gauss theorem [46–48].

2.3 Boundary Integral Equation Formulations

To derive the direct BIE corresponding to PDE (2.1), we apply the second Green's identity given in Eq. (1.24):

$$\int_V \left[u\nabla^2 v - v\nabla^2 u \right] dV = \int_S \left[u\frac{\partial v}{\partial n} - v\frac{\partial u}{\partial n} \right] dS.$$

(2.11)

Let $v(\mathbf{y}) = \phi(\mathbf{y})$, which satisfies Eq. (2.1), and $u(\mathbf{y}) = G(\mathbf{x}, \mathbf{y})$, which satisfies Eq. (2.4). We have, from identity Eq. (2.11):

$$\int_V \left[G(\mathbf{x}, \mathbf{y})\nabla^2\phi(\mathbf{y}) - \phi(\mathbf{y})\nabla^2 G(\mathbf{x}, \mathbf{y}) \right] dV(\mathbf{y})$$
$$= \int_S \left[G(\mathbf{x}, \mathbf{y})\frac{\partial\phi(\mathbf{y})}{\partial n(\mathbf{y})} - \phi(\mathbf{y})\frac{\partial G(\mathbf{x}, \mathbf{y})}{\partial n(\mathbf{y})} \right] dS(\mathbf{y}).$$

Applying Eqs. (2.1), (2.4), and (1.25), we obtain:

$$\phi(\mathbf{x}) = \int_S [G(\mathbf{x}, \mathbf{y})q(\mathbf{y}) - F(\mathbf{x}, \mathbf{y})\phi(\mathbf{y})] \, dS(\mathbf{y})$$
$$+ \int_V G(\mathbf{x}, \mathbf{y}) f(\mathbf{y}) dV(\mathbf{y}), \quad \forall \mathbf{x} \in V,$$

(2.12)

where $q = \partial\phi/\partial n$.

Equation (2.12) is the representation integral of the solution ϕ inside the domain V for Eq. (2.1). Once the boundary values of both ϕ and q are known on S, Eq. (2.12) can be applied to calculate ϕ everywhere in V, if needed.

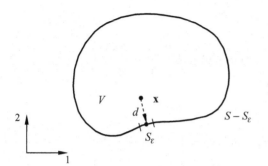

Figure 2.2. Limits as $\mathbf{x}$ approaches boundary S.

To solve the unknown boundary values of ϕ and q on S, we let $\mathbf{x}$ tend to S to obtain a BIE from Eq. (2.12). To do this, we consider the following limit:

$$\lim_{\mathbf{x}\to S}\phi(\mathbf{x}) = \lim_{\mathbf{x}\to S}\left\{\int_S [G(\mathbf{x}, \mathbf{y})q(\mathbf{y}) - F(\mathbf{x}, \mathbf{y})\phi(\mathbf{y})]\,dS(\mathbf{y}) + \int_V G(\mathbf{x}, \mathbf{y})f(\mathbf{y})dV(\mathbf{y})\right\}.$$
$$(2.13)$$

The kernel $G(\mathbf{x}, \mathbf{y})$ is *weakly singular* at $r = 0$ [of $O(\log r)$ in two dimensions and $O(1/r)$ in three dimensions] and $F(\mathbf{x}, \mathbf{y})$ is *strongly singular* [of $O(1/r)$ in two dimensions and $O(1/r^2)$ in three dimensions]. Therefore, we cannot place $\mathbf{x}$ on boundary S directly in Eq. (2.13). Careful consideration of the limit process is necessary for each integral on the right-hand side of Eq. (2.13).

We now proceed to use the 2D case as an example to see how to evaluate the limits in (2.13). We first divide the boundary S into two parts: $S - S_\varepsilon$ and S_ε, where S_ε is a small segment with length 2ε centered around the point to which $\mathbf{x}$ will approach (Figure 2.2).

The first integral on the right-hand side of (2.13) is evaluated as:

$$\lim_{\mathbf{x}\to S}\int_S G(\mathbf{x}, \mathbf{y})q(\mathbf{y})dS(\mathbf{y}) = \lim_{\varepsilon\to 0}\int_{S-S_\varepsilon} G(\mathbf{x}, \mathbf{y})q(\mathbf{y})dS(\mathbf{y})$$
$$+ \lim_{\substack{d\to 0 \\ \varepsilon\to 0}}\int_{S_\varepsilon} G(\mathbf{x}, \mathbf{y})dS(\mathbf{y})q(\mathbf{y}_\xi),$$

where $\mathbf{y}_\xi$ is a point on S_ε. When ε is small, S_ε can be regarded as a straight-line segment (assuming S is smooth); the analytical integration of G kernel on this line segment is given in Appendix A.1, Eq. (A.5). When Eq. (A.5) is used, the limit of this integral turns out to be:

$$\lim_{\substack{d\to 0 \\ \varepsilon\to 0}}\int_{S_\varepsilon} G(\mathbf{x}, \mathbf{y})dS(\mathbf{y})q(\mathbf{y}_\xi) = 0.$$

Therefore:

$$\lim_{\mathbf{x}\to S}\int_S G(\mathbf{x}, \mathbf{y})q(\mathbf{y})dS(\mathbf{y}) = \lim_{\varepsilon\to 0}\int_{S-S_\varepsilon} G(\mathbf{x}, \mathbf{y})q(\mathbf{y})dS(\mathbf{y}) = \int_S G(\mathbf{x}, \mathbf{y})q(\mathbf{y})dS(\mathbf{y}),$$
$$(2.14)$$

where the last integral is evaluated with the definition of a CPV integral. (For simplicity of notation, no special symbol is used here to indicate this fact.)

Similarly, the second integral on the right-hand side of (2.13) is evaluated as:

$$\lim_{\mathbf{x} \to S} \int_S F(\mathbf{x}, \mathbf{y})\phi(\mathbf{y})dS(\mathbf{y}) = \lim_{\varepsilon \to 0} \int_{S-S_\varepsilon} F(\mathbf{x}, \mathbf{y})\phi(\mathbf{y})dS(\mathbf{y})$$
$$+ \lim_{\substack{d \to 0 \\ \varepsilon \to 0}} \int_{S_\varepsilon} F(\mathbf{x}, \mathbf{y})dS(\mathbf{y})\phi(\mathbf{y}_\xi).$$

Applying the result in Eq. (A.6) of Appendix A.1, we obtain:

$$\lim_{\substack{d \to 0 \\ \varepsilon \to 0}} \int_{S_\varepsilon} F(\mathbf{x}, \mathbf{y})dS(\mathbf{y})\phi(\mathbf{y}_\xi) = -\frac{1}{2}\phi(\mathbf{x}), \quad \mathbf{x} \in S,$$

$$\lim_{\mathbf{x} \to S} \int_S F(\mathbf{x}, \mathbf{y})\phi(\mathbf{y})dS(\mathbf{y}) = \lim_{\varepsilon \to 0} \int_{S-S_\varepsilon} F(\mathbf{x}, \mathbf{y})\phi(\mathbf{y})dS(\mathbf{y}) - \frac{1}{2}\phi(\mathbf{x})$$
$$= \int_S F(\mathbf{x}, \mathbf{y})\phi(\mathbf{y})dS(\mathbf{y}) - \frac{1}{2}\phi(\mathbf{x}), \quad \mathbf{x} \in S, \quad (2.15)$$

where the last integral is understood as a CPV integral that is evaluated on $S - S_\varepsilon$ with $\varepsilon \to 0$. We see that there is a jump term associated with the integral with the F kernel as $\mathbf{x}$ approaches S. The third integral on the right-hand side of (2.13) has no jump term; that is:

$$\lim_{\mathbf{x} \to S} \int_V G(\mathbf{x}, \mathbf{y}) f(\mathbf{y})dV(\mathbf{y}) = \int_V G(\mathbf{x}, \mathbf{y}) f(\mathbf{y})dV(\mathbf{y}). \quad (2.16)$$

Substituting Eqs. (2.14)–(2.16) into (2.13) and combining the free terms, we arrive at the following *conventional* BIE (CBIE):

$$c(\mathbf{x})\phi(\mathbf{x}) = \int_S [G(\mathbf{x}, \mathbf{y})q(\mathbf{y}) - F(\mathbf{x}, \mathbf{y})\phi(\mathbf{y})] dS(\mathbf{y})$$
$$+ \int_V G(\mathbf{x}, \mathbf{y}) f(\mathbf{y})dV(\mathbf{y}), \quad \forall \mathbf{x} \in S, \quad (2.17)$$

in which $c(\mathbf{x})$ is a coefficient and $c(\mathbf{x}) = 1/2$ if S is smooth around $\mathbf{x}$. The same result can be derived for the 3D case. In this equation, both variables ϕ and q are now on the boundary S. Later, we will see that we can write CBIE (2.17) in a weakly singular form by using the integral identities for the fundamental solution, so that we do not need to evaluate the CPV integral (with the F kernel) and the constant $c(\mathbf{x})$ explicitly in the solutions of the BIE.

Treatment of the domain integral in CBIE (2.17) is discussed in Section 2.9 for the case in which $f(\mathbf{y})$ is nonzero over a finite area or volume within the domain V. When $f(\mathbf{y})$ is due to a concentrated or point source within V, we can write $f(\mathbf{y})$ as:

$$f(\mathbf{y}) = Q\delta(\mathbf{x}_Q, \mathbf{y}), \quad (2.18)$$

where $\mathbf{x}_Q$ is the location of the source and Q represents the intensity of the source. Using the sifting property of the Dirac δ function [Eq. (1.25)], we can evaluate the domain integral in CBIE (2.17) for this case as follows:

$$\int_V G(\mathbf{x}, \mathbf{y}) f(\mathbf{y}) dV(\mathbf{y}) = Q \int_V G(\mathbf{x}, \mathbf{y}) \delta(\mathbf{x}_Q, \mathbf{y}) dV(\mathbf{y}) = QG(\mathbf{x}, \mathbf{x}_Q). \quad (2.19)$$

This contribution is added to the right-hand side vector $\mathbf{b}$ of the BEM system of equations based on the CBIE (discussed in Section 2.5).

Once we obtain the unknown variables ϕ and q on S from solving CBIE (2.17), we can evaluate the potential inside the domain V by using the representation integral of (2.12), if needed. To evaluate the derivatives of the potential in V, we take the derivative of (2.12) to obtain:

$$\frac{\partial \phi}{\partial x_i}(\mathbf{x}) = \int_S \left[\frac{\partial G(\mathbf{x}, \mathbf{y})}{\partial x_i} q(\mathbf{y}) - \frac{\partial F(\mathbf{x}, \mathbf{y})}{\partial x_i} \phi(\mathbf{y}) \right] dS(\mathbf{y})$$
$$+ \int_V \frac{\partial G(\mathbf{x}, \mathbf{y})}{\partial x_i} f(\mathbf{y}) dV(\mathbf{y}), \quad \forall \mathbf{x} \in V. \quad (2.20)$$

Letting the source point $\mathbf{x}$ tend to boundary S and multiplying both sides of (2.20) with the normal at $\mathbf{x}$, we obtain the so called *hypersingular* BIE (HBIE):

$$c(\mathbf{x})q(\mathbf{x}) = \int_S \left[K(\mathbf{x}, \mathbf{y})q(\mathbf{y}) - H(\mathbf{x}, \mathbf{y})\phi(\mathbf{y}) \right] dS(\mathbf{y})$$
$$+ \int_V K(\mathbf{x}, \mathbf{y}) f(\mathbf{y}) dV(\mathbf{y}), \quad \forall \mathbf{x} \in S, \quad (2.21)$$

where the two new kernels are:

$$K(\mathbf{x}, \mathbf{y}) \equiv \frac{\partial G(\mathbf{x}, \mathbf{y})}{\partial n(\mathbf{x})} = \begin{cases} \dfrac{1}{2\pi r} r_{,k} n_k(\mathbf{x}), & \text{for two dimensions,} \\[2mm] \dfrac{1}{4\pi r^2} r_{,k} n_k(\mathbf{x}), & \text{for three dimensions,} \end{cases} \quad (2.22)$$

$$H(\mathbf{x}, \mathbf{y}) \equiv \frac{\partial F(\mathbf{x}, \mathbf{y})}{\partial n(\mathbf{x})}$$
$$= \begin{cases} \dfrac{1}{2\pi r^2} \left[n_k(\mathbf{x})n_k(\mathbf{y}) - 2r_{,k} n_k(\mathbf{x})r_{,l} n_l(\mathbf{y}) \right], & \text{for two dimensions} \\[2mm] \dfrac{1}{4\pi r^3} \left[n_k(\mathbf{x})n_k(\mathbf{y}) - 3r_{,k} n_k(\mathbf{x})r_{,l} n_l(\mathbf{y}) \right], & \text{for three dimensions} \end{cases}.$$
$$(2.23)$$

$K(\mathbf{x}, \mathbf{y})$ kernel is strongly singular, and the first integral in HBIE (2.21) is a CPV integral, the $H(\mathbf{x}, \mathbf{y})$ kernel is hypersingular, and the second integral in (2.21) is a HFP integral. HBIE (2.21) also can be written in a weakly singular form, and we do not need to evaluate these singular or hypersingular integrals in the BEM unless they can be evaluated readily (as in the constant-element case that we discuss in the next section).

The CBIE degenerates when it is applied to solve crack problems or thin inclusion problems [49]. In these cases, the HBIE can be applied alone or in combination with the CBIE to have a nondegenerate dual BIE formulation for crack problems and thin-shape problems. We will see some examples later in this and subsequent chapters.

CBIE (2.17) and HBIE (2.21) are also valid for an *infinite domain problem*, where the domain is outside a closed boundary S and extends to infinity. We can show that contributions of integrals on the boundaries at infinity vanish if we assume that $\phi(R) \sim O(1/R^{\alpha})$ and $q(R) \sim O(1/R^{1+\alpha})$, as $R \to \infty$, where R is the radius of a large circle (2D) or sphere (3D) and the real number $\alpha > 0$.

2.4 Weakly Singular Forms of the Boundary Integral Equations

CBIE (2.17) and HBIE (2.21) can be recast into forms that involve only weakly singular integrals [46–48] or even nonsingular forms without any singular integrals [47]. For example, using the first identity in (2.7) for the fundamental solution $G(\mathbf{x}, \mathbf{y})$, we can show that the coefficient $c(\mathbf{x})$ in CBIE (2.17) can be written as:

$$c(\mathbf{x}) = 1 + \lim_{\substack{d \to 0 \\ \varepsilon \to 0}} \int_{S_\varepsilon} F(\mathbf{x}, \mathbf{y})dS(\mathbf{y}) = \gamma - \lim_{\varepsilon \to 0} \int_{S-S_\varepsilon} F(\mathbf{x}, \mathbf{y})dS(\mathbf{y})$$

$$= \gamma - \int_S F(\mathbf{x}, \mathbf{y})dS(\mathbf{y}), \quad \forall \mathbf{x} \in S \quad \text{(a CPV integral)}, \qquad (2.24)$$

in which $\gamma = 0$ for finite domain and $\gamma = 1$ for infinite domain problems. Substituting the preceding expression for $c(\mathbf{x})$ in CBIE (2.17), we obtain the following *weakly singular form* of the CBIE:

$$\gamma\phi(\mathbf{x}) + \int_S F(\mathbf{x}, \mathbf{y}) \left[\phi(\mathbf{y}) - \phi(\mathbf{x})\right] dS(\mathbf{y})$$

$$= \int_S G(\mathbf{x}, \mathbf{y})q(\mathbf{y})dS(\mathbf{y}) + \int_V G(\mathbf{x}, \mathbf{y}) f(\mathbf{y})dV(\mathbf{y}), \quad \forall \mathbf{x} \in S, \qquad (2.25)$$

in which the integral with the F kernel is now weakly singular, because

$$F(\mathbf{x}, \mathbf{y}) \left[\phi(\mathbf{y}) - \phi(\mathbf{x})\right] \sim \begin{cases} O\left(\dfrac{1}{r}\right) O(r) = O(1), & \text{for two dimensions,} \\[2ex] O\left(\dfrac{1}{r^2}\right) O(r) = O\left(\dfrac{1}{r}\right), & \text{for three dimensions,} \end{cases}$$

as $r \to 0$, if ϕ is continuous. Similarly, using the first three identities for the fundamental solution, we can derive the following *weakly singular form* of the

HBIE (see Refs. [50, 51] for the results of a Helmholtz equation with a Laplace equation as a special case):

$$\gamma q(\mathbf{x}) + \int_S H(\mathbf{x}, \mathbf{y}) \left[\phi(\mathbf{y}) - \phi(\mathbf{x}) - \frac{\partial \phi}{\partial \xi_\alpha}(\mathbf{x})(\xi_\alpha - \xi_{o\alpha}) \right] dS(\mathbf{y})$$

$$+ e_{\alpha k} \frac{\partial \phi}{\partial \xi_\alpha}(\mathbf{x}) \int_S \left[K(\mathbf{x}, \mathbf{y}) n_k(\mathbf{y}) + F(\mathbf{x}, \mathbf{y}) n_k(\mathbf{x}) \right] dS(\mathbf{y})$$

$$= \int_S \left[K(\mathbf{x}, \mathbf{y}) + F(\mathbf{x}, \mathbf{y}) \right] q(\mathbf{y}) \, dS(\mathbf{y})$$

$$- \int_S F(\mathbf{x}, \mathbf{y}) \left[q(\mathbf{y}) - q(\mathbf{x}) \right] dS(\mathbf{y}) + \int_V K(\mathbf{x}, \mathbf{y}) f(\mathbf{y}) dV(\mathbf{y}), \quad \forall \mathbf{x} \in S,$$

$$(2.26)$$

in which ξ_α and $\xi_{o\alpha}$ are the coordinates of $\mathbf{y}$ and $\mathbf{x}$, respectively, in tangential directions ($\alpha = 1$ for two dimensions and $\alpha = 1, 2$ for three dimensions) in the local (natural) coordinate system on an element and $e_{\alpha k} = \partial \xi_\alpha / \partial x_k$ [51]. All the integrals in (2.26) are now, at most, weakly singular if ϕ has continuous first derivatives.

Weakly singular forms of the BIEs, or *regularized* BIEs, which do not contain any strongly singular and hypersingular integrals, are useful in cases in which higher-order boundary elements are applied to solve the BIEs. In these cases, analytical evaluations of the singular integrals are difficult or impossible to obtain and the use of numerical integrations is troublesome. When constant elements are used, all the singular and hypersingular integrals can be evaluated analytically (see Appendix A.1); therefore, the original singular forms of CBIE (2.17) and HBIE (2.21) can be applied directly.

2.5 Discretization of the Boundary Integral Equations for 2D Problems Using Constant Elements

We now apply the boundary elements to discretize the BIEs in order to solve them numerically for the unknown boundary variables. As an example, we discretize the CBIE (assuming $f = 0$) in (2.17) for 2D problems by using *constant* elements.

First, we divide the boundary S into line segments (elements) ΔS_j and place one node on each element (Figure 2.3). The total number of elements is M, and the total number of nodes is N. In the case of using constant elements, we have $N = M$. Next, we place the source point $\mathbf{x}$ at node i and notice that:

$$\phi(\mathbf{y}) = \phi_j, \quad q(\mathbf{y}) = q_j, \quad \text{on element } \Delta S_j,$$

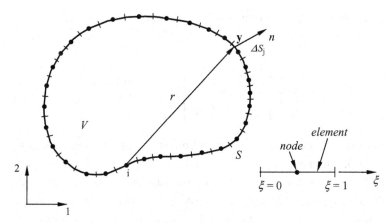

Figure 2.3. Discretization of boundary S using *constant* elements.

where ϕ_j and q_j ($j = 1, 2, \ldots, N$) are the nodal values of ϕ and q, respectively, on element ΔS_j for constant elements. CBIE (2.17) becomes:

$$\frac{1}{2}\phi_i = \sum_{j=1}^{N} \int_{\Delta S_j} [G_i q_j - F_i \phi_j] \, dS = \sum_{j=1}^{N} \left[\int_{\Delta S_j} G_i \, dS q_j - \int_{\Delta S_j} F_i \, dS \phi_j \right], \quad (2.27)$$

where G_i and F_i are the kernels with the source point $\mathbf{x}$ placed at node i. We obtain the following discretized equation of CBIE (2.17) for node i:

$$\frac{1}{2}\phi_i = \sum_{j=1}^{N} [g_{ij} q_j - \hat{f}_{ij} \phi_j], \quad \text{for } i = 1, 2, \ldots, N, \quad (2.28)$$

where the coefficients are given by:

$$g_{ij} = \int_{\Delta S_j} G_i \, dS, \qquad \hat{f}_{ij} = \int_{\Delta S_j} F_i \, dS, \quad \text{for } i, j = 1, 2, \ldots, N. \quad (2.29)$$

The preceding integrals can be evaluated *analytically* for all singular ($i = j$) or nonsingular ($i \neq j$) cases with the constant elements (see Appendix A.1).

In matrix form, Eq. (2.28) can be written as:

$$\begin{bmatrix} f_{11} & f_{12} & \cdots & f_{1N} \\ f_{21} & f_{22} & \cdots & f_{2N} \\ \vdots & \vdots & \ddots & \vdots \\ f_{N1} & f_{N2} & \cdots & f_{NN} \end{bmatrix} \begin{Bmatrix} \phi_1 \\ \phi_2 \\ \vdots \\ \phi_N \end{Bmatrix} = \begin{bmatrix} g_{11} & g_{12} & \cdots & g_{1N} \\ g_{21} & g_{22} & \cdots & g_{2N} \\ \vdots & \vdots & \ddots & \vdots \\ g_{N1} & g_{N2} & \cdots & g_{NN} \end{bmatrix} \begin{Bmatrix} q_1 \\ q_2 \\ \vdots \\ q_N \end{Bmatrix}, \quad (2.30)$$

where $f_{ij} = \hat{f}_{ij} + \frac{1}{2}\delta_{ij}$. In the conventional BEM approach, we form a standard linear system of equations as follows by applying the boundary condition at each node and switching the columns in the two matrices in Eq. (2.30):

$$
\begin{bmatrix}
a_{11} & a_{12} & \cdots & a_{1N} \\
a_{21} & a_{22} & \cdots & a_{2N} \\
\vdots & \vdots & \ddots & \vdots \\
a_{N1} & a_{N2} & \cdots & a_{NN}
\end{bmatrix}
\begin{Bmatrix}
\lambda_1 \\
\lambda_2 \\
\vdots \\
\lambda_N
\end{Bmatrix}
=
\begin{Bmatrix}
b_1 \\
b_2 \\
\vdots \\
b_N
\end{Bmatrix}, \quad \text{or} \quad \mathbf{A}\boldsymbol{\lambda} = \mathbf{b}, \qquad (2.31)
$$

where $\mathbf{A}$ is the coefficient matrix, $\boldsymbol{\lambda}$ is the unknown vector (with unknown ϕ or q at each node), and $\mathbf{b}$ is the known right-hand-side vector. Obviously, the construction of matrix $\mathbf{A}$ requires $O(N^2)$ operations using Eqs. (2.29), and the size of the required memory for storing $\mathbf{A}$ is also $O(N^2)$ because $\mathbf{A}$ is, in general, a nonsymmetric and dense matrix. The solution of the system in Eq. (2.31) using direct solvers such as Gauss elimination requires $O(N^3)$ operations because of this general matrix. Thus, the conventional BEM approach by solving Eq. (2.31) directly can handle only BEM models with a few thousand equations on a desktop computer with 1-GB RAM (GB is gigabyte and RAM is random-access memory).

By solving Eq. (2.31), we can obtain all the unknown boundary variables on each element. If the fields inside the domain are demanded, we can compute ϕ by using integral representation (2.12) and the derivatives of ϕ by using (2.20) in similar discretized forms. Discretization of the BIEs using constant elements is straightforward, and all the integrals of the kernels on the elements can be evaluated analytically. However, the accuracy of the constant elements is not very good, and usually more constant elements are needed to obtain reasonably accurate BEM results as compared with those obtained with high-order elements.

2.6 Using Higher-Order Elements

Higher-order boundary elements are needed to improve the accuracy and efficiency of the BEM solutions in situations in which accuracy and efficiency are critical, such as stress concentration problems. For curved boundaries, higher-order elements, such as quadratic elements, are also beneficial because of the more accurate representation of the geometry. However, the use of higher-order elements also presents some challenges. Analytical integrations of the coefficients are no longer available, in general, and numerical integrations need to be used. In the following subsection, we discuss the linear and quadratic elements for 2D problems.

2.6.1 Linear Elements

For discretization using linear elements (Figure 2.4), each element is associated with two nodes placed at the ends of the element. The element is assumed

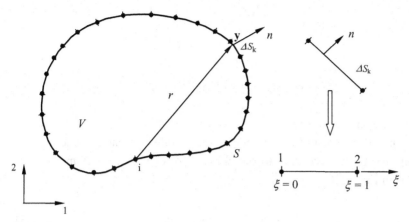

Figure 2.4. Discretization of boundary S using *linear* elements.

to be straight, and the fields are assumed to vary linearly over the element. Two shape functions are introduced to represent the function on an element. For example, on element ΔS_k ($k = 1, 2, 3, \ldots, M$, with M being the total number of elements), we have:

$$\phi(\mathbf{y}) = \phi(\xi) = \sum_{\alpha=1}^{2} N_\alpha(\xi)\phi^\alpha, \tag{2.32}$$

$$q(\mathbf{y}) = q(\xi) = \sum_{\alpha=1}^{2} N_\alpha(\xi)q^\alpha, \tag{2.33}$$

where ϕ^1, ϕ^2 and q^1, q^2 are the nodal values of ϕ and q at local nodes 1 and 2, respectively; ξ is the local (natural) coordinate defined on the element; and $N_1(\xi)$ and $N_2(\xi)$ are the linear shape functions given by:

$$N_1(\xi) = 1 - \xi, \quad \text{and} \quad N_2(\xi) = \xi. \tag{2.34}$$

Placing source point $\mathbf{x}$ at node i ($i = 1, 2, 3, \ldots, N$), we have the following discretized equation for CBIE (2.17) (with $f = 0$):

$$c_i\phi_i = \sum_{k=1}^{M} \int_{\Delta S_k} [G_i q - F_i \phi] \, dS,$$

$$= \sum_{k=1}^{M} \int_{\Delta S_k} \left[G_i \sum_{\alpha=1}^{2} N_\alpha q^\alpha \right] dS - \sum_{k=1}^{M} \int_{\Delta S_k} \left[F_i \sum_{\alpha=1}^{2} N_\alpha \phi^\alpha \right] dS,$$

$$= \sum_{k=1}^{M} \sum_{\alpha=1}^{2} \left[\int_{\Delta S_k} G_i N_\alpha dS \right] q^\alpha - \sum_{k=1}^{M} \sum_{\alpha=1}^{2} \left[\int_{\Delta S_k} F_i N_\alpha dS \right] \phi^\alpha, \tag{2.35}$$

that is,

$$c_i\phi_i = \sum_{k=1}^{M} \sum_{\alpha=1}^{2} g_{ik}^\alpha q^\alpha - \sum_{k=1}^{M} \sum_{\alpha=1}^{2} \hat{f}_{ik}^\alpha \phi^\alpha, \tag{2.36}$$

where:

$$g_{ik}^{\alpha} = \int_{\Delta S_k} G_i N_\alpha dS$$

$$\hat{f}_{ik}^{\alpha} = \int_{\Delta S_k} F_i N_\alpha dS \qquad (2.37)$$

with $i = 1, 2, 3, \ldots, N$ (number of nodes), $k = 1, 2, 3, \ldots, M$ (number of elements), and $\alpha = 1$ and 2 (number of local nodes on each element). Rearranging the terms according to the global nodes (instead of elements), we obtain from Eq. (2.36):

$$c_i \phi_i = \sum_{j=1}^{N} g_{ij} q_j - \sum_{j=1}^{N} \hat{f}_{ij} \phi_j, \qquad (2.38)$$

where g_{ij} and $\hat{f}_{ij}$ are sums of the integrals g_{ik}^{α} and $\hat{f}_{ik}^{\alpha}$ on elements around node j, respectively. Thus, we have a linear system of equation similar to Eq. (2.28) and the matrix form is identical to Eq. (2.30), where $f_{ij} = \hat{f}_{ij} + c_i \delta_{ij}$ (no sum over i).

In general, numerical integration schemes need to be used to evaluate the coefficients in (2.30) using formulas (2.37). For example, for nondiagonal terms ($i \neq j$), we have:

$$g_{ik}^{\alpha} = \int_{\Delta S_k} G_i N_\alpha dS = \int_0^1 G_i[\mathbf{x}, \mathbf{y}(\xi)] N_\alpha(\xi) |J| d\xi, \qquad (2.39)$$

where the global coordinate $\mathbf{y}$ is related to the local coordinate by:

$$y_l(\xi) = \sum_{\alpha=1}^{2} N_\alpha(\xi) y_l^{\alpha}, \quad \text{for } l = 1, 2,$$

with y_l^{α} being the nodal values of y_l, and:

$$dS = \sqrt{\left(\frac{dy_1}{d\xi}\right)^2 + \left(\frac{dy_2}{d\xi}\right)^2} d\xi = |J| d\xi, \quad \text{where} \quad |J| = \sqrt{\left(\frac{dy_1}{d\xi}\right)^2 + \left(\frac{dy_2}{d\xi}\right)^2}$$

is the Jacobian of the coordinate transformation. The integral on the right-hand side of Eq. (2.39) can be evaluated by standard Gaussian quadrature. In most cases, a four-point quadrature should be sufficient. The second integral in (2.37) is handled in a similar way.

For the diagonal terms, we can evaluate the coefficients analytically by using the definition of CPV integrals. The results are:

$$g_{ii} = \frac{L_a}{8\pi} \left[3 + 2\log\left(\frac{1}{L_a}\right)\right] + \frac{L_b}{8\pi} \left[3 + 2\log\left(\frac{1}{L_b}\right)\right],$$

$$\hat{f}_{ii} = 0, \quad i = 1, 2, 3, \ldots, N; \qquad (2.40)$$

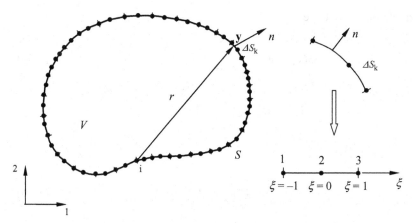

Figure 2.5. Discretization of boundary S using *quadratic* elements.

in which L_a and L_b are the lengths of the two elements before and after node i. For coefficient f_{ii}, there is an easy way to calculate their values. Suppose we have a uniform potential field, with $\phi = 1$ and $q = 0$ everywhere. Then, from Eq. (2.30), we obtain:

$$f_{ii} = -\sum_{j\neq i}^{N} f_{ij}, \tag{2.41}$$

for *finite* domain problems, which avoids calculation of c_i at each node. For *infinite* domain problems, the contributions from integrals at infinity do not vanish for uniform potentials. Thus, the relation in (2.41) is changed to:

$$f_{ii} = 1 - \sum_{j\neq i}^{N} f_{ij}. \tag{2.42}$$

Results in (2.41) and (2.42) are exact, meaning that there is no additional error introduced. We can derive these results analytically by using identity (2.7) for the fundamental solution [46].

2.6.2 Quadratic Elements

Quadratic elements can be used for problems demanding even higher accuracy, such as problems with singular fields caused by cracklike objects, problems with curved boundaries, and so on. There are three nodes on a quadratic element (Figure 2.5). The element can be a quadratic curve, which is a more accurate representation for a domain with curved boundaries. The three quadratic-shape functions are given as follows in the local coordinate ξ:

$$N_1(\xi) = \frac{1}{2}\xi(\xi - 1), \quad N_2(\xi) = (1 - \xi)(1 + \xi), \quad N_3(\xi) = \frac{1}{2}\xi(\xi + 1). \tag{2.43}$$

On each element, we have:

$$\phi(\mathbf{y}) = \phi(\xi) = \sum_{\alpha=1}^{3} N_\alpha(\xi)\phi^\alpha, \tag{2.44}$$

$$q(\mathbf{y}) = q(\xi) = \sum_{\alpha=1}^{3} N_\alpha(\xi)q^\alpha, \tag{2.45}$$

and for the geometry:

$$y_l(\xi) = \sum_{\alpha=1}^{3} N_\alpha(\xi)y_l^\alpha, \quad \text{for } l = 1, 2. \tag{2.46}$$

Using quadratic elements, we can write the discretized form of CBIE (2.17) (with $f = 0$) as:

$$c_i\phi_i = \sum_{k=1}^{M}\sum_{\alpha=1}^{3} g_{ik}^\alpha q^\alpha - \sum_{k=1}^{M}\sum_{\alpha=1}^{3} \hat{f}_{ik}^\alpha \phi^\alpha, \tag{2.47}$$

in which:

$$\begin{aligned} g_{ik}^\alpha &= \int_{\Delta S_k} G_i N_\alpha dS, \\ \hat{f}_{ik}^\alpha &= \int_{\Delta S_k} F_i N_\alpha dS, \end{aligned} \tag{2.48}$$

with $i = 1, 2, 3, \ldots, N$, $k = 1, 2, 3, \ldots, M$, and $\alpha = 1, 2$, and 3. Rearranging the terms according to the global nodes (based on the element connectivity information), we obtain a system of equations similar to that given in (2.30). In this case, all the coefficients g_{ij} and f_{ij} in (2.30) need to be calculated numerically by Gaussian quadrature, except for f_{ii}, which still can be determined by Eq. (2.41) for a finite domain or Eq. (2.42) for an infinite domain, without introducing any additional errors.

2.7 Discretization of the Boundary Integral Equations for 3D Problems

For 3D problems, surfaces of a domain will be discretized using surface elements, which can be constant, linear, or quadratic (Figure 2.6). The shape of an element can be triangular or quadrilateral. For a constant element, there is only one node located at the center of the element. For a linear element, there is one node at each vertex of the element. For a quadratic element, there is one node at each vertex and on each edge of the element. Implementation of the constant elements is straightforward, and analytical integrations of the kernels

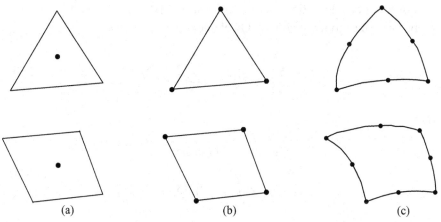

Figure 2.6. Surface elements for 3D problems: (a) constant, (b) linear, (c) quadratic.

are possible. However, using linear and quadratic elements is more accurate and efficient.

We use the quadrilateral four-node (Q4) linear elements (Figure 2.7) as an example to see how to discretize the CBIE for 3D problems.

In the natural coordinate system (ξ, η), the four shape functions are:

$$
\begin{aligned}
N_1(\xi, \eta) &= \frac{1}{4}(1 - \xi)(1 - \eta), \\
N_2(\xi, \eta) &= \frac{1}{4}(1 + \xi)(1 - \eta), \\
N_3(\xi, \eta) &= \frac{1}{4}(1 + \xi)(1 + \eta), \\
N_4(\xi, \eta) &= \frac{1}{4}(1 - \xi)(1 + \eta).
\end{aligned}
\tag{2.49}
$$

Note that $\sum_{\alpha=1}^{4} N_\alpha = 1$ at any point inside the element, as expected.

Figure 2.7. A Q4 linear element for 3D problems.

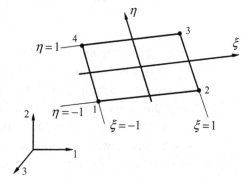

For a 3D problem, the discretized form of CBIE (2.17) (with $f = 0$) can still be written as follows with the Q4 elements:

$$c_i \phi_i = \sum_{k=1}^{M} \sum_{\alpha=1}^{4} g_{ik}^{\alpha} q^{\alpha} - \sum_{k=1}^{M} \sum_{\alpha=1}^{4} \hat{f}_{ik}^{\alpha} \phi^{\alpha}, \qquad (2.50)$$

where:

$$g_{ik}^{\alpha} = \int_{\Delta S_k} G_i N_\alpha dS,$$
$$\hat{f}_{ik}^{\alpha} = \int_{\Delta S_k} F_i N_\alpha dS, \qquad (2.51)$$

with $i = 1, 2, 3, \ldots, N$ (total number of nodes on the surface), $k = 1, 2, 3, \ldots,$ M (total number of elements), and $\alpha = 1, 2, 3$ and 4 (for Q4 elements). Information about the element connectivity is needed to assemble the system of equations as given in Eq. (2.30). For example, if the global node number of local node α of element k is j, then coefficient g_{ik}^{α} should go to the ith row and jth column of the g matrix on the right-hand side of Eq. (2.30).

In 3D cases, all the coefficients g_{ij} and f_{ij} we determine by using (2.51) are surface integrals that we can calculate numerically using Gaussian quadrature, except for f_{ii}, which we can still determine by Eq. (2.41) for a finite domain or Eq. (2.42) for an infinite domain. For example, to compute g_{ij}, we proceed as follows:

$$g_{ik}^{\alpha} = \int_{\Delta S_k} G_i N_\alpha dS = \int_{-1}^{1} \int_{-1}^{1} G_i [\mathbf{x}, \mathbf{y}(\xi, \eta)] N_\alpha(\xi, \eta) |\mathbf{J}| d\xi \, d\eta, \qquad (2.52)$$

where $|\mathbf{J}|$ is the determinant of the Jacobian matrix; that is:

$$|\mathbf{J}| = \det \begin{bmatrix} \hat{i}_1 & \hat{i}_2 & \hat{i}_3 \\ \dfrac{\partial y_1}{\partial \xi} & \dfrac{\partial y_2}{\partial \xi} & \dfrac{\partial y_3}{\partial \xi} \\ \dfrac{\partial y_1}{\partial \eta} & \dfrac{\partial y_2}{\partial \eta} & \dfrac{\partial y_3}{\partial \eta} \end{bmatrix}, \qquad (2.53)$$

with $\hat{i}_k$ being the unit base vector along the y_k axis.

Quadrilateral eight-node (Q8) quadratic elements (Figure 2.8) have also been used widely in the BEM for 3D problems because of their accuracy and flexibility in modeling curved surfaces. Using quadratic elements is even more beneficial for the conventional BEM because they can deliver more accurate results with fewer elements when the number of elements is limited by the method or the computer.

Figure 2.8. A Q8 quadratic element for 3D problems.

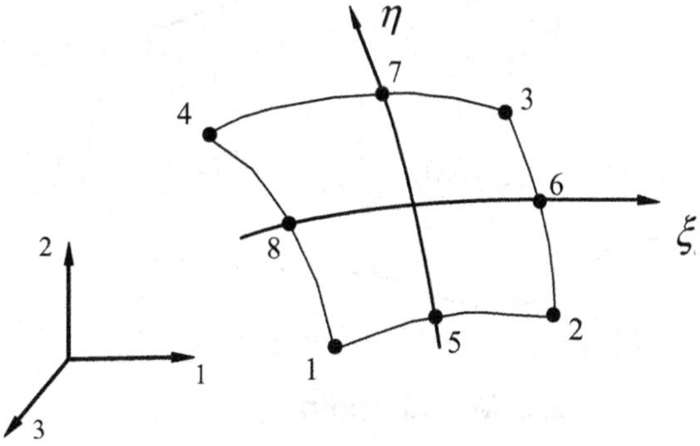

In the natural coordinate system (ξ, η), the eight shape functions for Q8 elements are:

$$N_1(\xi, \eta) = \frac{1}{4}(1 - \xi)(\eta - 1)(\xi + \eta + 1),$$

$$N_2(\xi, \eta) = \frac{1}{4}(1 + \xi)(\eta - 1)(\eta - \xi + 1),$$

$$N_3(\xi, \eta) = \frac{1}{4}(1 + \xi)(1 + \eta)(\xi + \eta - 1),$$

$$N_4(\xi, \eta) = \frac{1}{4}(\xi - 1)(\eta + 1)(\xi - \eta + 1),$$

$$N_5(\xi, \eta) = \frac{1}{2}(1 - \eta)(1 - \xi^2),$$

$$N_6(\xi, \eta) = \frac{1}{2}(1 + \xi)(1 - \eta^2),$$

$$N_7(\xi, \eta) = \frac{1}{2}(1 + \eta)(1 - \xi^2),$$

$$N_8(\xi, \eta) = \frac{1}{2}(1 - \xi)(1 - \eta^2).$$

(2.54)

Again, we have the relation $\sum_{\alpha=1}^{8} N_\alpha = 1$ at any point (ξ, η) inside the element. Both the physical fields and the geometry (coordinates) are interpolated using these shape functions in a manner similar to that previously discussed for Q4 elements.

It is difficult to evaluate analytically the singular (CPV) and hypersingular (HFP) integrals on 3D curved elements. Therefore, the weakly singular forms of the BIEs can be applied to avoid direct evaluation of such singular integrals, especially using any numerical integration scheme. Treatment of various singular integrals using the weakly singular forms of the BIEs and quadratic surface elements can be found in Refs. [18, 46, 50–54].

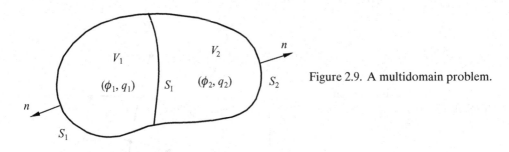

Figure 2.9. A multidomain problem.

2.8 Multidomain Problems

For multidomain problems, regions of different materials can be treated separately first by the BEM and then assembled together with the interface conditions. For example, suppose that we need to solve the potential problem in a multiple domain comprising two material regions V_1 and V_2 with the interface S_I (Figure 2.9).

For region 1, we have the following BEM equation from Eq. (2.30):

$$\begin{bmatrix} \mathbf{F}_1 & \mathbf{F}_1^I \end{bmatrix} \begin{Bmatrix} \boldsymbol{\phi}_1 \\ \boldsymbol{\phi}_1^I \end{Bmatrix} = \begin{bmatrix} \mathbf{G}_1 & \mathbf{G}_1^I \end{bmatrix} \begin{Bmatrix} \mathbf{q}_1 \\ \mathbf{q}_1^I \end{Bmatrix}, \tag{2.55}$$

and for region 2:

$$\begin{bmatrix} \mathbf{F}_2 & \mathbf{F}_2^I \end{bmatrix} \begin{Bmatrix} \boldsymbol{\phi}_2 \\ \boldsymbol{\phi}_2^I \end{Bmatrix} = \begin{bmatrix} \mathbf{G}_2 & \mathbf{G}_2^I \end{bmatrix} \begin{Bmatrix} \mathbf{q}_2 \\ \mathbf{q}_2^I \end{Bmatrix}, \tag{2.56}$$

where f_m and $\mathbf{q}_m$ are the nodal values of ϕ and q, respectively, on the boundary S_m of domain m; and $\boldsymbol{\phi}_m^I$ and $\mathbf{q}_m^I$ are the nodal values of ϕ and q, respectively, on interface S_I from domain m (here, $m = 1$ or 2). Applying the interface conditions (assuming perfect bonding):

$$\begin{aligned} \boldsymbol{\phi}_1^I &= \boldsymbol{\phi}_2^I \equiv \boldsymbol{\phi}^I, \\ \mathbf{q}_1^I &= -\mathbf{q}_2^I \equiv \mathbf{q}^I, \end{aligned} \tag{2.57}$$

we can write Eqs. (2.55) and (2.56) in a single matrix equation as:

$$\begin{bmatrix} \mathbf{F}_1 & \mathbf{F}_1^I & \mathbf{0} \\ \mathbf{0} & \mathbf{F}_2^I & \mathbf{F}_2 \end{bmatrix} \begin{Bmatrix} \boldsymbol{\phi}_1 \\ \boldsymbol{\phi}^I \\ \boldsymbol{\phi}_2 \end{Bmatrix} = \begin{bmatrix} \mathbf{G}_1 & \mathbf{G}_1^I & \mathbf{0} \\ \mathbf{0} & -\mathbf{G}_2^I & \mathbf{G}_2 \end{bmatrix} \begin{Bmatrix} \mathbf{q}_1 \\ \mathbf{q}^I \\ \mathbf{q}_2 \end{Bmatrix}. \tag{2.58}$$

Moving the unknown term $\mathbf{q}^I$ to the left-hand side, we obtain:

$$\begin{bmatrix} \mathbf{F}_1 & \mathbf{0} & \mathbf{F}_1^I & -\mathbf{G}_1^I \\ \mathbf{0} & \mathbf{F}_2 & \mathbf{F}_2^I & \mathbf{G}_2^I \end{bmatrix} \begin{Bmatrix} \boldsymbol{\phi}_1 \\ \boldsymbol{\phi}_2 \\ \boldsymbol{\phi}^I \\ \mathbf{q}^I \end{Bmatrix} = \begin{bmatrix} \mathbf{G}_1 & \mathbf{0} \\ \mathbf{0} & \mathbf{G}_2 \end{bmatrix} \begin{Bmatrix} \mathbf{q}_1 \\ \mathbf{q}_2 \end{Bmatrix}. \tag{2.59}$$

It is noticed that for multidomain problems, the matrices of the BEM system of equations become banded, which will be more obvious when more sub-domains are involved. This is an advantage for solving the system of equations because of the improved conditioning. For problems with slender domains, even if they are not multidomain problems, we can apply the multidomain technique to reduce the bandwidth of the equations.

2.9 Treatment of the Domain Integrals

If the function $f(\mathbf{x})$ in CBIE (2.17) is not zero over a finite area or volume, we need to deal with this domain integral that contains no unknown variables. There are several options in evaluations of the domain integrals. Some basic approaches are reviewed briefly in the following subsections. More advanced techniques for dealing with various domain integrals in the BEM can be found in the literature, such as the dual reciprocal methods (see, e.g., Ref. [55]).

2.9.1 Numerical Integration Using Internal Cells

In this approach, we simply divide the domain V into L cells V_k ($k = 1, 2, \ldots, L$) and proceed as follows, with the source point $\mathbf{x}$ placed at node i on the boundary:

$$b_i = \int_V G(\mathbf{x}, \mathbf{y}) f(\mathbf{y}) dV(\mathbf{y}) = \int_V G_i f dV = \sum_{k=1}^{L} \int_{V_k} G_i f dV, \qquad (2.60)$$

where the integral on each cell can be evaluated numerically with a Gaussian quadrature. The contribution b_i is added to the right-hand side vector in Eq. (2.31). The internal cells can be coarser and do not need to match the mesh on the boundary. This is the easiest and earliest approach for dealing with the domain integrals in the BEM. However, it is no longer used widely because of the need to use the domain cells, which is not consistent with the boundary approach.

2.9.2 Transformation to Boundary Integrals

A more elegant approach to deal with domain integrals is to transform them into boundary integrals and use the same boundary mesh to evaluate them as that used to solve the boundary variables. There were many methods developed in the past 30 years or so in this regard, including the dual reciprocal methods. A very basic method is given here as an example.

Suppose that $f(\mathbf{x})$ is a harmonic function (e.g., f is constant or linear over the domain V); we have $\nabla^2 f = 0$. Next, we write the fundamental solution as:

$$G(\mathbf{x}, \mathbf{y}) = \nabla^2 G^*(\mathbf{x}, \mathbf{y}). \tag{2.61}$$

This is possible because for two dimensions, we have:

$$G^*(\mathbf{x}, \mathbf{y}) = \frac{1}{8\pi} \left[\log\left(\frac{1}{r}\right) + 1 \right] r^2, \tag{2.62}$$

and for three dimensions, we have:

$$G^*(\mathbf{x}, \mathbf{y}) = \frac{1}{8\pi} r. \tag{2.63}$$

Applying the Green's second identity (2.11), we evaluate:

$$\int_V G(\mathbf{x}, \mathbf{y}) f(\mathbf{y}) dV(\mathbf{y}) = \int_V \left(\nabla^2 G^* \right) f dV$$
$$= \int_V G^* \left(\nabla^2 f \right) dV + \int_S \left(\frac{\partial G^*}{\partial n} f - G^* \frac{\partial f}{\partial n} \right) dS;$$

that is,

$$\int_V G(\mathbf{x}, \mathbf{y}) f(\mathbf{y}) dV(\mathbf{y}) = \int_S \left(\frac{\partial G^*}{\partial n} f - G^* \frac{\partial f}{\partial n} \right) dS, \tag{2.64}$$

which transforms the domain integral into a boundary integral.

2.9.3 Use of Particular Solutions

In this approach, we simply seek to find a particular solution ϕ^p of Eq. (2.1), such that $\phi = \phi^c + \phi^p$ and:

$$\nabla^2 \phi^p + f = 0, \quad \nabla^2 \phi^c = 0. \tag{2.65}$$

Thus, the free term f is taken care of by the particular solution ϕ^p. The problem is reduced to solving a Laplace equation for ϕ^c under modified boundary conditions.

2.10 Indirect Boundary Integral Equation Formulations

We can use the fundamental solutions to construct BIEs directly, without using the Green's identities. The BIEs constructed in this way often contain density functions that do not have direct physical meanings. Thus, these BIEs are called *indirect BIE formulations*. For example, consider the following integral representation:

$$\phi(\mathbf{x}) = \int_S G(\mathbf{x}, \mathbf{y}) \sigma(\mathbf{y}) dS(\mathbf{y}), \quad \forall \mathbf{x} \in V, \tag{2.66}$$

which is called a *single-layer potential* [56]. It can be shown that $\phi(\mathbf{x})$ given by (2.66) satisfies the Laplace equation [Eq. (2.1) with $f = 0$]. The density function $\sigma(\mathbf{y})$ has no clear physical meaning in this case. Field $\phi(\mathbf{x})$ can be determined by Eq. (2.66) after the density function $\sigma(\mathbf{y})$ is found on the boundary. Taking the derivative of (2.66) and letting $\mathbf{x}$ approach boundary S, we can obtain the following two BIEs:

$$\phi(\mathbf{x}) = \int_S G(\mathbf{x}, \mathbf{y})\sigma(\mathbf{y})dS(\mathbf{y}), \quad \forall \mathbf{x} \in S, \tag{2.67}$$

$$\frac{\partial \phi}{\partial n}(\mathbf{x}) = \int_S \frac{\partial G(\mathbf{x}, \mathbf{y})}{\partial n(\mathbf{x})}\sigma(\mathbf{y})dS(\mathbf{y}) + \frac{1}{2}\sigma(\mathbf{x}), \quad \forall \mathbf{x} \in S. \tag{2.68}$$

If we use (2.67) on S_ϕ where ϕ is given (Dirichlet BC), we obtain from (2.67):

$$\overline{\phi}(\mathbf{x}) = \int_S G(\mathbf{x}, \mathbf{y})\sigma(\mathbf{y})dS(\mathbf{y}), \quad \forall \mathbf{x} \in S, \tag{2.69}$$

which is an integral equation of the *first* kind. If we use (2.68) on S_q where q is given (Neumann BC), we obtain from (2.68):

$$\overline{q}(\mathbf{x}) = \int_S \frac{\partial G(\mathbf{x}, \mathbf{y})}{\partial n(\mathbf{x})}\sigma(\mathbf{y})dS(\mathbf{y}) + \frac{1}{2}\sigma(\mathbf{x}), \quad \forall \mathbf{x} \in S, \tag{2.70}$$

which is an integral equation of the *second* kind. BEM equations based on Eqs. (2.67) and (2.68) can be applied to solve for unknown density $\sigma(\mathbf{y})$ over the entire boundary S. Then, going back to the single-layer potential representation of $\phi(\mathbf{x})$ in Eq. (2.66), we can evaluate $\phi(\mathbf{x})$ everywhere inside the domain V. This is one of the indirect BIE formulations in the BEM.

Starting with the following *double-layer potential* [56] representation:

$$\phi(\mathbf{x}) = \int_S \frac{\partial G(\mathbf{x}, \mathbf{y})}{\partial n(\mathbf{y})}\mu(\mathbf{y})dS(\mathbf{y}), \quad \forall \mathbf{x} \in V, \tag{2.71}$$

we can formulate another indirect BIE formulation for potential problems.

The advantages of using indirect BIE formulations are that fewer integrals need to be computed to form the BEM system of equations, and better conditioning of the BEM equations can be achieved by selecting integral equations of the second kind based on the boundary conditions. The disadvantage of using indirect BIEs is obvious, in that the density functions $\sigma(\mathbf{y})$ and $\mu(\mathbf{y})$ are not the physical quantities directly, and a postprocessing step is needed to obtain the field $\phi(\mathbf{x})$, which can offset the savings in forming the BIE equations. In addition, better conditioning can always be achieved with a combination of the CBIE and HBIE (direct BIEs), as discussed later.

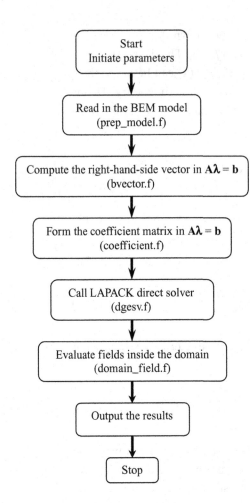

Figure 2.10. Flowchart for a conventional BEM program for solving 2D potential problems.

2.11 Programming for the Conventional Boundary Element Method

A sample program written in Fortran for solving general 2D potential problems is provided in Appendix B.1. In the conventional BEM approach, we need to first form the BEM system of equations, as shown in Eq. (2.31), then use a direct solver (e.g., using Gauss elimination) or an iterative solver (e.g., the generalized minimal residual method [GMRES]) to solve the linear systems, and finally we evaluate the field inside the domain if needed.

The flowchart as shown in Figure 2.10 is typical for conventional BEM programs using the Fortran language. The main components of a BEM program are subroutines for reading the model data (nodes, elements, BCs, and field points inside the domain), computing the right-hand-side vector $\mathbf{b}$, computing the coefficients to form the system matrix $\mathbf{A}$, solving the system of equations (in this case, using the direct solver from LAPACK), evaluating the field values inside the domain when needed, and writing the output files.

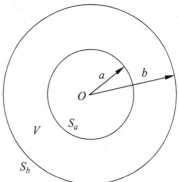

Figure 2.11. A simple potential problem in an annular region V.

The subroutines indicated in the flowchart in Figure 2.10 are those given in Appendix B.1 for the program for solving general 2D potential problems with the CBIE and using constant elements. Programs for other problems or using other types of elements may need a few additional subroutines, such as those for numerical integration of singular and nonsingular integrals.

For a beginner in BEM research and development, it is very important and beneficial if he or she can write a 2D BEM code using the conventional BEM approach first, so that he or she can understand the structure of a BEM program and implementation of its major components. The program provided in Appendix B.1 can serve as a starting point. Explanations of the main program, subroutines, and main variables are provided within the code. A sample input file is also provided in Appendix B.3, which can be used as a template to prepare input files for other 2D potential problems. In this program, all the integrals, including the singular ones, are computed with the analytical results given in Appendix A.1, which is possible only with constant elements.

2.12 Numerical Examples

A few examples are given in this section to show the accuracy of the BEM for solving potential problems. More examples, especially those involving large-scale problems, are given in the next chapter related to the fast multipole BEM.

2.12.1 An Annular Region

We first consider a simple potential (e.g., heat conduction) problem in a 2D annular region, as shown in Figure 2.11, for which the available analytical solution can be used to verify the BEM results. The BEM program given in Appendix B.1 is used in this study. The field ϕ is given on the inner boundary S_a, and the normal derivative is given on the outer boundary S_b.

Table 2.1. *Results of the potential and*
normal derivative for the annular region

N	q_a	ϕ_b
36	−401.7715	376.7236
72	−400.4007	377.1410
360	−400.0148	377.2548
720	−400.0036	377.2579
1440	−400.0005	377.2586
2400	−400.0006	377.2588
4800	−400.0006	377.2589
7200	−399.9982	377.2589
9600	−399.9969	377.2589
Analytical solution	−400.0000	377.2589

The analytical solution for this (axisymmetric) problem is given by:

$$\phi(r) = \phi_a + q_b b \log\left(\frac{r}{a}\right), \tag{2.72}$$

where ϕ_a and q_b are the given values of ϕ and q on boundaries S_a and S_b, respectively, and r is the radial coordinate in a polar-coordinate system centered at O. This gives:

$$\phi_b = \phi(b) = \phi_a + q_b b \log\left(\frac{b}{a}\right), \quad q_a = \frac{\partial\phi}{\partial n}(a) = -q_b\frac{b}{a}. \tag{2.73}$$

For this problem, we choose $a = 1, b = 2, \phi_a = 100$, and $q_b = 200$. This gives:

$$\phi_b = 377.258872, \quad q_a = -400.0.$$

We discretize the inner and outer boundaries with the same number of elements. Table 2.1 shows the results of ϕ_b and q_a for this problem as the total number of elements increases from 36 to 9600. As we can see, the results for the BEM converge quickly to the exact solution for the mesh with only 72 constant elements with a relative error of 0.1%. The results continue to improve until reaching the mesh with 4800 elements. For the two larger meshes (with 7200 and 9600 elements), the results for q_a deviate slightly from the exact solution, which may be caused by numerical errors that are due to the extremely small elements in the mesh.

2.12.2 Electrostatic Fields Outside Two Conducting Beams

Next, we consider the electrostatic field surrounding two thin beams that are applied with opposite voltages (Figure 2.12). This is an exterior problem that is also governed by the Laplace equation. However, for this problem, a *dual BIE formulation* that is a linear combination of CBIE (2.17) and HBIE (2.21)

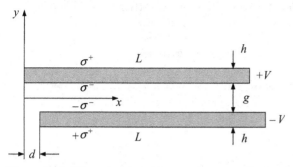

Figure 2.12. Electrostatic field around two parallel beams.

is used to overcome the difficulties associated with thin shapes if the CBIE is applied alone [49, 57].

In Figure 2.12, the length of the beam is L, the thickness is h, and the gap between the two beams is g. An offset d in the x direction also may be introduced between the two beams. A potential (voltage) V is applied to the top beam, and the negative potential $(-V)$ is applied to the bottom beam. For this problem, the analytical solution for the charge density σ^- on the lower surface of the top beam is given by (see, e.g., Ref. [58]):

$$\sigma^- = \varepsilon \frac{\partial \phi}{\partial n} = \varepsilon \frac{\Delta V}{\Delta n} = \varepsilon \frac{2V}{g} \tag{2.74}$$

for the region away from the edges of the beams. This formula is used to verify the BEM results.

The parameters used here are $\varepsilon = 1$, $L = 0.01$ m, $h = 0.0001$ m, $g = 0.0011$ m, $d = 0$, and $V = 1$. Constant elements are used. The number of elements along the beam-length direction is increased from 10, 20, 50, to 100, and 5 elements are used on each edge (side) of the beams, corresponding to BEM models with 30, 50, 110, and 210 elements per beam, respectively. The BEM results obtained with the dual BIE formulations converge very quickly. Figure 2.13 shows the convergence of the BEM results for the charge densities on the lower and upper surfaces of the top beam. In fact, the model with just 10 elements along the beam-length direction yields a value of σ^- at the middle of the lower surface of the top beam that agrees with the analytical solution ($\sigma^- = 1818$ in this case) within the first four digits.

Figure 2.14 shows the charge density on the top beam in the same parallel beam model but with an offset $d = g = 0.0011$ m and using 210 elements per beam. The charge densities in the middle of the beam remain the same ($\sigma^- = 1818$), whereas the fields near the edges have marked changes. The charge densities on the bottom beam have negative values and are "antisymmetrical" relative to the results on the top beam and thus are not plotted.

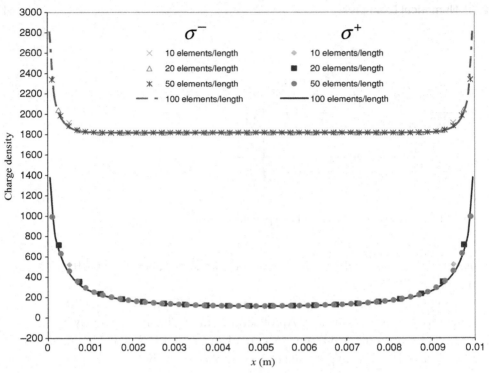

Figure 2.13. Convergence of the BEM results obtained using a dual BIE on the top beam in the parallel beam model ($\varepsilon = 1$, $L = 0.01$ m, $h = 0.0001$ m, $g = 0.0011$ m, $d = 0$, and $V = 1$).

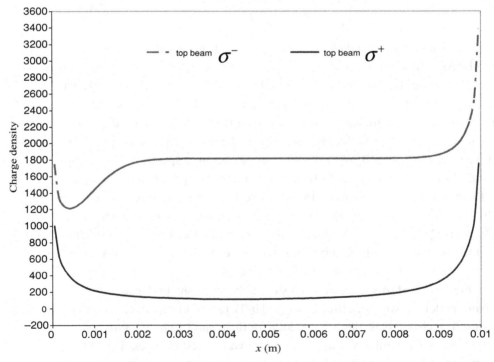

Figure 2.14. Charge density on the top beam in the parallel beam model with offset ($\varepsilon = 1$, $L = 0.01$ m, $h = 0.0001$ m, $g = 0.0011$ m, $d = g$, $V = 1$, and 210 elements per beam).

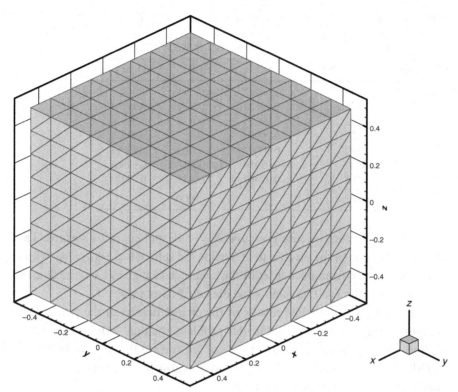

Figure 2.15. A cube meshed with 768 elements and with linear potential in the x direction.

2.12.3 Potential Field in a Cube

Two 3D examples are given next using a BEM program with constant surface elements [59]. A cube is shown in Figure 2.15, which is a simple interior problem used to show the accuracy of the 3D code with constant elements. The cube has an edge length $= 1$ and is applied with a linear potential $\phi(x, y, z) = x$ on all surfaces. The normal derivative q for this problem should be 1 on the surface at $x = 0.5$ and -1 on the surface at $x = -0.5$.

Table 2.2 shows the results obtained with the conventional BEM and by using the CBIE, HBIE, and dual BIE for BEM meshes with increasing numbers of elements. One can conclude from these results that the HBIE and the dual BIE are equally as accurate as the CBIE. Note that constant triangular elements are used in this study. If linear or quadratic elements were applied, a few elements should have been sufficient for obtaining results with a similar level of accuracy because of the specified linear field.

2.12.4 Electrostatic Field Outside a Conducting Sphere

A single conducting sphere model (Figure 2.16) is shown next. This is a simple exterior problem with curved boundaries. The conducting sphere has a radius

Table 2.2. *Results for the cube with a linear potential in the x direction*

Model		Normal derivative at $(0.5, 0, 0)$		
Elem/edge	Total DOFs	CBIE	HBIE	Dual BIE
2	48	1.08953	1.07225	1.06800
4	192	0.99124	1.00624	0.99754
8	768	0.99825	1.00438	0.99894
12	1728	0.99908	1.00327	0.99934
16	3072	0.99942	1.00260	0.99953
20	4800	0.99959	1.00216	0.99963
24	6912	0.99969	1.00185	0.99970
	Exact Value		1.00000	

$a = 1$, and a constant electric potential $\phi_0 = 1$ is applied on its surface. The analytical solution of the electric field outside the sphere is $\phi = (a/r)\phi_0$, with r being the distance from the center of the sphere, which gives a charge density on the surface equal to 1, assuming the dielectric constant $\varepsilon = 1$.

Table 2.3 gives the BEM results of the charge density at the point $(1, 0, 0)$ on the surface of the sphere. For this problem, the dual BIE is slightly less accurate than the CBIE because of the curved surface that cannot be represented accurately by constant elements and can cause the evaluations of hypersingular integrals to be less accurate [59].

Several numerical examples for solving both 2D and 3D potential problems are presented in this subsection. Constant elements are used for all the examples, and reasonably accurate BEM solutions are obtained. Linear or quadratic elements can be applied to improve the accuracy of the BEM solutions (see problems). These examples are used again in the next chapter on

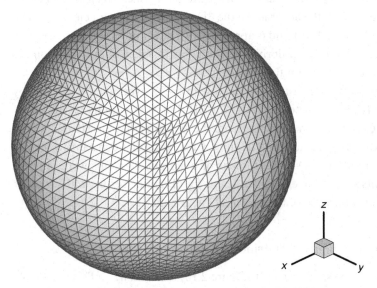

Figure 2.16. A spherical perfect conductor meshed with 4800 elements.

Table 2.3. *Results for the single perfect conducting sphere*

Model DOFs	Charge density at $(1, 0, 0)$	
	CBIE	Dual BIE
768	0.98749	0.95086
1728	0.99377	0.96609
3072	0.99634	0.97431
4800	0.99761	0.97937
6912	0.99832	0.98278
Exact Value	1.00000	

the fast multipole solution techniques to demonstrate the computational efficiencies of the fast multipole BEM for solving large-scale problems.

2.13 Summary

In this chapter, the BIE formulations for solving potential problems are presented. It is shown that the partial differential equation (Poisson equation or Laplace equation) can be transformed into BIEs with the help of the fundamental solution and the Green's identity. Both the conventional BIE and the hypersingular BIE formulations are discussed. Weakly singular forms of these BIEs are also presented to show that singular integrals in the BIE formulations and therefore their BEM solutions can be avoided altogether if the integral terms are arranged properly. The discretization procedures are discussed with constant, linear, and quadratic line elements for 2D problems and with linear and quadratic surface elements for 3D problems. Programming for the BEM using the conventional approach is discussed briefly, and several numerical examples are presented.

This chapter is the basis for all other chapters dealing with fast multipole solution techniques for potential, elasticity, Stokes flow, and acoustic wave problems. The basic ideas, BIE formulations, BEM discretization procedures, programming, and solutions for those problems are similar to these discussed in this chapter. Therefore, it is very important to understand all of the material covered in this chapter before moving on to the following chapters.

Problems

2.1. Show that $G(\mathbf{x}, \mathbf{y})$ given by Eq. (2.5) does satisfy Eq. (2.4); that is, $\nabla^2 G(\mathbf{x}, \mathbf{y}) = 0$, for $r \neq 0$; and near $r = 0$, $-\nabla^2 G(\mathbf{x}, \mathbf{y})$ behaves like a $\delta(\mathbf{x}, \mathbf{y})$ function. For example, $-\int_{V_\varepsilon} \nabla^2 G(\mathbf{x}, \mathbf{y}) dV(\mathbf{y}) = 1$, where V_ε is a circular region centered at $\mathbf{x}$ with radius ε.

2.2. Verify that ϕ given by integral representation (2.12) does satisfy Poisson equation (2.1).

2.3. Show that:

$$r_{,ij} = \frac{1}{r}\left(\delta_{ij} - r_{,i}\,r_{,j}\right), \tag{2.75}$$

where $r = \sqrt{(y_i - x_i)(y_i - x_i)}$ is the distance between source point $\mathbf{x}$ and field point $\mathbf{y}$.

2.4. Verify the second integral identity for the fundamental solution $G(\mathbf{x}, \mathbf{y})$ given in (2.8).

2.5. Show that CBIE (2.17) and HBIE (2.21) are also valid for infinite domain problems; that is, contributions of integrals on boundaries at infinity should vanish.

2.6. Verify the weakly singular form of the CBIE in (2.25) using integral identity (2.7).

2.7. Verify formulas in Eqs. (2.40) for linear elements.

2.8. Show that the result in Eq. (2.42) is true for infinite domain problems regardless of which type of element is used.

2.9. Applying the program in Appendix B.1, solve the cylinder problem shown in Figure 2.11 by using a quarter-symmetry model and compare your results with those presented in Table 2.1.

2.10. Develop a program (in Fortran, C/C++, or Matlab) using *linear* line elements for solving general 2D potential problems. You can start with the program using constant elements given in Appendix B.1.

2.11. Develop a program (in Fortran, C/C++, or Matlab) using *quadratic* line elements for solving general 2D potential problems. You can start with the program using constant elements given in Appendix B.1.

3 Fast Multipole Boundary Element Method for Potential Problems

Although the BEM has enjoyed the reputation of ease in modeling or meshing for problems with complicated geometries or in infinite domains, its efficiency in solutions has been a serious drawback for analyzing large-scale models. For example, the BEM has been limited to solving problems with only a few thousand DOFs on a PC for many years. This is because the conventional BEM, as described in the previous chapter, produces dense and nonsymmetric matrices that although smaller in sizes, require $O(N^2)$ operations for computing the coefficients and $O(N^3)$ operations for solving the system by using direct solvers (N is the number of equations of the linear system or DOFs).

In the mid-1980s, Rokhlin and Greengard [33–35] pioneered the innovative fast multipole method (FMM) that can be used to accelerate the solutions of BEM by severalfold, promising to reduce the CPU time in FMM-accelerated BEM to $O(N)$. With the help of the FMM, the BEM can now solve large-scale problems that are beyond the reach of other methods. We call the fast multipole accelerated BEM *fast multipole BEM* or simply *fast BEM* from now on to distinguish it from the conventional BEM described in the previous chapter. Some of the early work on fast multipole BEMs in mechanics can be found in Refs. [36–40], which show the great promise of the fast multipole BEM for solving large-scale problems. A comprehensive review of the fast multipole BIE/BEM research up to 2002 can be found in Ref. [41].

In this chapter, the FMM for solving the BEM systems of equations for potential problems is introduced. First, the fast multipole BEM for 2D potential problems is discussed in detail. Then, the fast multipole BEM for 3D potential problems is introduced. Several examples of modeling large-scale potential problems are provided. This chapter forms the basis for all subsequent chapters on fast multipole BEM approaches for elasticity, Stokes flow, and acoustic wave problems.

3.1 Basic Ideas in the Fast Multipole Method

To facilitate the discussion, the BEM system of equations (2.30) is repeated here:

$$\begin{bmatrix} f_{11} & f_{12} & \cdots & f_{1N} \\ f_{21} & f_{22} & \cdots & f_{2N} \\ \vdots & \vdots & \ddots & \vdots \\ f_{N1} & f_{N2} & \cdots & f_{NN} \end{bmatrix} \begin{Bmatrix} \phi_1 \\ \phi_2 \\ \vdots \\ \phi_N \end{Bmatrix} = \begin{bmatrix} g_{11} & g_{12} & \cdots & g_{1N} \\ g_{21} & g_{22} & \cdots & g_{2N} \\ \vdots & \vdots & \ddots & \vdots \\ g_{N1} & g_{N2} & \cdots & g_{NN} \end{bmatrix} \begin{Bmatrix} q_1 \\ q_2 \\ \vdots \\ q_N \end{Bmatrix}. \quad (3.1)$$

After the boundary conditions are applied, a standard linear system of equations [Eq. (2.31)] is formed as follows by switching the columns in the two matrices in Eq. (3.1):

$$\begin{bmatrix} a_{11} & a_{12} & \cdots & a_{1N} \\ a_{21} & a_{22} & \cdots & a_{2N} \\ \vdots & \vdots & \ddots & \vdots \\ a_{N1} & a_{N2} & \cdots & a_{NN} \end{bmatrix} \begin{Bmatrix} \lambda_1 \\ \lambda_2 \\ \vdots \\ \lambda_N \end{Bmatrix} = \begin{Bmatrix} b_1 \\ b_2 \\ \vdots \\ b_N \end{Bmatrix}, \quad \text{or} \quad \mathbf{A}\lambda = \mathbf{b}, \quad (3.2)$$

where $\mathbf{A}$ is the coefficient matrix, λ is the unknown vector, and $\mathbf{b}$ is the known right-hand-side vector. Obviously, the construction of matrix $\mathbf{A}$ requires $O(N^2)$ operations and the size of the required memory for storing $\mathbf{A}$ is also $O(N^2)$ because $\mathbf{A}$ is, in general, a nonsymmetric and dense matrix. The solution of the system in Eq. (3.2) by use of direct solvers such as Gauss elimination is even worse, requiring $O(N^3)$ operations because of this general matrix. Even with iterative solvers, the solution time is still $O(N^2)$. That is why the conventional BEM approach for solving BIEs is, in general, slow and inefficient for large-scale problems despite its robustness in the meshing stage as compared with other domain-based methods.

The main idea of the fast multipole BEM is to apply iterative solvers (e.g., GMRES) to solve Eq. (3.2) and use the FMM to accelerate the matrix–vector multiplication ($\mathbf{A}\lambda$) in each iteration, without ever forming the entire matrix $\mathbf{A}$ explicitly. Direct integrations are still needed when the elements are close to the source point, whereas fast multipole expansions are used for elements that are far away from the source point. Figure 3.1 is a graphical illustration of the fast multipole BEM compared with the conventional BEM. For the far-field calculations, the node-to-node (or element-to-element) interactions in the conventional BEM [Figure 3.1(a)] are replaced with cell-to-cell interactions [Figure 3.1(b)] by a hierarchical tree structure of cells containing groups of elements (in Figure 3.1, the dots indicate nodes and cells and the lines indicate the interactions needed). This is possible by use of the multipole and local expansions of the integrals and some translations that are discussed in the following section. The numbers of lines represent the computational complexities

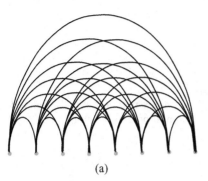
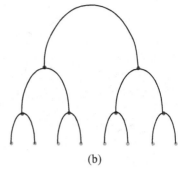

(a)   (b)

Figure 3.1. A graphical illustration of (a) the conventional BEM approach $[O(N^2)]$, and (b) the fast multipole BEM $[O(N)$ for $\log N]$.

of the two approaches, and a dramatic decrease of operations in the fast multipole BEM is obvious from this illustration.

A fundamental reason for the reduction in operations in the fast multipole BEM, as shown in Figure 3.1(b), is due to the fact that the Green's functions or the kernels in the BIEs can be expanded in the following form:

$$G(\mathbf{x}, \mathbf{y}) = \sum_i G_i^x(\mathbf{x}, \mathbf{y}_c) G_i^y(\mathbf{y}, \mathbf{y}_c), \qquad (3.3)$$

where $\mathbf{y}_c$ is an expansion point. This can be achieved by use of various forms of expansions, including but not limited to Taylor series expansions. By using an expansion as in Eq. (3.3), we can write the original integral, such as the one with the G kernel in CBIE (2.17), as:

$$\int_{S_c} G(\mathbf{x}, \mathbf{y}) q(\mathbf{y}) dS(\mathbf{y}) = \sum_i G_i^x(\mathbf{x}, \mathbf{y}_c) \int_{S_c} G_i^y(\mathbf{y}, \mathbf{y}_c) q(\mathbf{y}) dS(\mathbf{y}), \qquad (3.4)$$

where S_c is a subset of S away from $\mathbf{x}$. In the conventional BEM, the integral is computed with the expression on the left-hand side of Eq. (3.4) directly. Any changes in the location of the source point $\mathbf{x}$ will require reevaluation of the entire integral. In the fast multipole BEM, when the source point $\mathbf{x}$ is far away from S_c, the original integral is computed with the expression on the right-hand side of Eq. (3.4), in which the new integrals need to be evaluated only once, independent of the locations of the source point $\mathbf{x}$. That is, the direct relation between $\mathbf{x}$ and $\mathbf{y}$ is cut off by use of the expansion and introduction of the new "middle" point $\mathbf{y}_c$. Additional expansions and translations, as well as the hierarchical tree structure of the elements, are introduced in the fast multipole BEM to further reduce the computational costs.

Using the FMM for the BEM, we can reduce the solution time to $O(N)$ for large-scale problems [41]. We can also reduce the memory requirement to $O(N)$ because, with iterative solvers, the entire matrix does not need to be stored in the memory. This drastic improvement in computing efficiency has presented many opportunities for the BEM. Large BEM models with a couple

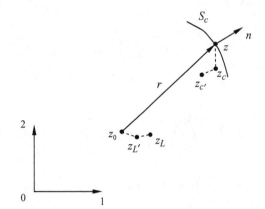

Figure 3.2. Complex notation and the related points for fast multipole expansions.

million DOFs that could not be solved by the conventional BEM before can now be solved readily by using the fast multipole BEM within hours on a PC or BEM models with tens of millions of DOFs on a supercomputer.

3.2 Fast Multipole Boundary Element Method for 2D Potential Problems

In this section, we first discuss the expansions that are used in the FMM for 2D potential problems. Then, the main procedures and algorithms in the fast multipole BEM are described.

We first consider the following integral with the G kernel in CBIE (2.17):

$$\int_{S_c} G(\mathbf{x}, \mathbf{y}) q(\mathbf{y}) dS(\mathbf{y}), \tag{3.5}$$

in which S_c is a subset of boundary S and away from the source point $\mathbf{x}$.

For convenience, we introduce *complex notation*; that is, we replace the source point:

$$\mathbf{x} \;\Rightarrow\; z_0 = x_1 + i x_2$$

and the field point:

$$\mathbf{y} \;\Rightarrow\; z = y_1 + i y_2$$

in the complex plane, where $i = \sqrt{-1}$ (Figure 3.2). Using the complex notation, we can write:

$$G(\mathbf{x}, \mathbf{y}) = \mathrm{Re}\{G(z_0, z)\}, \tag{3.6}$$

where:

$$G(z_0, z) = -\frac{1}{2\pi} \log(z_0 - z) \tag{3.7}$$

is the fundamental solution in complex notation and Re{ } indicates the real part of the variable or function. Thus, the integral in (3.5) is equivalent to the real part of the following integral:

$$\int_{S_c} G(z_0, z)q(z)dS(z),$$
(3.8)

where $q(z)$ is still a real-valued function of complex variable z.

We now introduce several important concepts in the FMM that form the building blocks for the fast multipole BEM.

3.2.1 Multipole Expansion (Moments)

The first idea is to expand the kernel function to see if we can separate the source point z_0 (**x**) and field point z (**y**). To do this, we introduce an expansion point z_c that is close to the field point z (Figure 3.2); that is, $|z - z_c| \ll |z_0 - z_c|$. We can write:

$$G(z_0, z) = -\frac{1}{2\pi} \log(z_0 - z) = -\frac{1}{2\pi} \left[\log(z_0 - z_c) + \log\left(1 - \frac{z - z_c}{z_0 - z_c} \right) \right].$$
(3.9)

Applying the following Taylor series expansion:

$$\log(1 - \xi) = -\sum_{k=1}^{\infty} \frac{\xi^k}{k}, \quad \text{for } |\xi| < 1,$$
(3.10)

to the second logarithmic term on the right-hand side of Eq. (3.9), we obtain:

$$G(z_0, z) = \frac{1}{2\pi} \sum_{k=0}^{\infty} O_k(z_0 - z_c) I_k(z - z_c).$$
(3.11)

We previously introduced two auxiliary functions $I_k(z)$ and $O_k(z)$ defined by:

$$I_k(z) = \frac{z^k}{k!}, \quad \text{for } k \geq 0;$$

$$O_k(z) = \frac{(k-1)!}{z^k}, \quad \text{for } k \geq 1; \quad \text{and} \quad O_0(z) = -\log(z).$$
(3.12)

The derivatives of functions $I_k(z)$ and $O_k(z)$ satisfy:

$$I'_k(z) = I_{k-1}(z), \quad \text{for } k \geq 1; \quad \text{and} \quad I'_0(z) = 0;$$

$$O'_k(z) = -O_{k+1}(z), \quad \text{for } k \geq 0.$$
(3.13)

In addition, we have the following two results:

$$I_k(z_1 + z_2) = \sum_{l=0}^{k} I_{k-l}(z_1) I_l(z_2) = \sum_{l=0}^{k} I_l(z_1) I_{k-l}(z_2);$$

$$O_k(z_1 + z_2) = \sum_{l=0}^{\infty} (-1)^l O_{k+l}(z_1) I_l(z_2), \quad \text{for } |z_2| < |z_1|.$$
(3.14)

The first equation is simply the binomial formula and the second is simply a Taylor series expansion of O_k about point z_1.

Note that in the G kernel given in Eq. (3.11), z_0 and z are now separated because of the introduction of the "middle point" z_c, which is a key in the FMM. The integral in (3.8) is now evaluated as follows:

$$\int_{S_c} G(z_0, z)q(z)dS(z) = \frac{1}{2\pi}\int_{S_c}\left[\sum_{k=0}^{\infty}O_k(z_0 - z_c)I_k(z - z_c)\right]q(z)dS(z);$$

that is, the *multipole expansion*:

$$\int_{S_c} G(z_0, z)q(z)dS(z) = \frac{1}{2\pi}\sum_{k=0}^{\infty}O_k(z_0 - z_c)M_k(z_c), \qquad (3.15)$$

where:

$$M_k(z_c) = \int_{S_c} I_k(z - z_c)q(z)dS(z), \quad k = 0, 1, 2, \ldots, \qquad (3.16)$$

are called *moments* about z_c, which are independent of the collocation point z_0 and need to be computed only once. After these moments are obtained, the G kernel integral can be evaluated readily by using Eq. (3.15) for any collocation point z_0 away from S_c (which will be within a *cell* centered at z_c).

We can evaluate the moments analytically by using the complex notation on constant elements. Suppose we have a line element starting at point z_a and ending at z_b. From Eq. (3.16), the contribution to the moment from this element can be evaluated as:

$$M_k^{(e)}(z_c) = \int_{z_a}^{z_b} I_k(z - z_c)q(z)dS(z) = q_e\int_{z_a}^{z_b} I_k(z - z_c)dS(z)$$

$$= q_e\int_{z_a}^{z_b}\frac{(z - z_c)^k}{k!}dS(z), \quad (3.17)$$

in which q_e is the nodal value of q on this element. Notice the relation:

$$dz = dy_1 + idy_2 = \left[\left(\frac{dy_1}{dS}\right) + i\left(\frac{dy_2}{dS}\right)\right]dS = \omega dS, \qquad (3.18)$$

where ω is the complex (unit) tangential vector along the boundary S. Using this relation, we can evaluate the preceding moment contribution as:

$$M_k^{(e)}(z_c) = q_e\int_{z_a}^{z_b}\frac{(z - z_c)^k}{k!}\overline{\omega}dz = q_e\overline{\omega}\left[I_{k+1}(z_b - z_c) - I_{k+1}(z_a - z_c)\right], \quad (3.19)$$

where $\overline{\omega}$ is the complex conjugate of ω. This analytical result can facilitate very efficient and accurate evaluations of the moments defined in Eq. (3.16) for constant elements.

3.2.2 Error Estimate for the Multipole Expansion

Errors in the multipole expansion are controlled by the number of terms used in the expansion in (3.11). An error bound can be derived readily for this multipole expansion (cf. results in Ref. [35]). If we apply a multipole expansion with p terms in Eq. (3.15), we have for the error bound:

$$E_M^p \equiv \left| \int_{S_c} G(z_0, z)q(z)dS(z) - \frac{1}{2\pi} \sum_{k=0}^{p} O_k(z_0 - z_c)M_k(z_c) \right|$$

$$= \frac{1}{2\pi} \left| \sum_{k=p+1}^{\infty} O_k(z_0 - z_c)M_k(z_c) \right|$$

$$\leq \frac{1}{2\pi} \sum_{k=p+1}^{\infty} |O_k(z_0 - z_c)| |M_k(z_c)|$$

$$\leq \frac{1}{2\pi} \sum_{k=p+1}^{\infty} |O_k(z_0 - z_c)| \left| \int_{S_c} I_k(z - z_c)q(z)dS(z) \right|$$

$$\leq \frac{1}{2\pi} \sum_{k=p+1}^{\infty} |O_k(z_0 - z_c)| \int_{S_c} |I_k(z - z_c)| |q(z)| dS(z)$$

$$\leq \frac{A}{2\pi} \sum_{k=p+1}^{\infty} |O_k(z_0 - z_c)| \frac{R^k}{k!}$$

$$= \frac{A}{2\pi} \sum_{k=p+1}^{\infty} \frac{(k-1)!}{|z_0 - z_c|^k} \frac{R^k}{k!} \leq \frac{A}{2\pi} \sum_{k=p+1}^{\infty} \frac{R^k}{|z_0 - z_c|^k}$$

$$= \frac{A}{2\pi} \frac{R^{p+1}}{|z_0 - z_c|^{p+1}} \frac{1}{1 - R/|z_0 - z_c|},$$

in which R is the radius of a region centered at z_c such that:

$$|z - z_c| < R \quad \text{and} \quad A \equiv \int_{S_c} |q(z)| dS(z). \tag{3.20}$$

Let $\rho = |z_0 - z_c|/R$; the preceding estimate of the error bound can be written as:

$$E_M^p \leq \frac{A}{2\pi} \frac{1}{(\rho - 1)} \left(\frac{1}{\rho} \right)^p. \tag{3.21}$$

We notice from estimate (3.21) that the larger the value of ρ, the smaller the value of this estimate of the error bound. If $\rho \geq 2$ – that is, when $|z_0 - z_c| \geq 2R$ – we have the following estimate:

$$E_M^p \leq \frac{A}{2\pi} \left(\frac{1}{2} \right)^p. \tag{3.22}$$

An error bound can be used to estimate the number (p) of the expansion terms so that it can be determined automatically by the computer program.

3.2.3 Moment-to-Moment Translation

If the expansion point z_c is moved to a new location $z_{c'}$ (Figure 3.2), we can apply a translation to obtain the moment at the new location without recomputing the moment by using Eq. (3.16). We obtain this translation by considering the following for the moments:

$$M_k(z_{c'}) = \int_{S_c} I_k(z - z_{c'})q(z)dS(z)$$

$$= \int_{S_c} I_k\left[(z - z_c) + (z_c - z_{c'})\right]q(z)dS(z).$$

Applying the binomial formula or the first equation in Eq. (3.14), we obtain:

$$M_k(z_{c'}) = \sum_{l=0}^{k} I_{k-l}(z_c - z_{c'})M_l(z_c). \tag{3.23}$$

This is the *moment-to-moment (M2M) translation* for the moments in which z_c is moved to $z_{c'}$. Note that there are only a finite number of terms needed in this translation; that is, no additional truncation error is introduced in M2M translations.

3.2.4 Local Expansion and Moment-to-Local Translation

Next, we introduce another expansion, the so-called local expansion about the source point z_0 (**x**). Suppose z_L is a point close to the source point z_0 (Figure 3.2); that is, $|z_0 - z_L| \ll |z_L - z_c|$. From the multipole expansion in Eq. (3.15), we have:

$$\int_{S_c} G(z_0, z)q(z)dS(z) = \frac{1}{2\pi}\sum_{k=0}^{\infty} O_k(z_0 - z_c)M_k(z_c)$$

$$= \frac{1}{2\pi}\sum_{k=0}^{\infty} O_k\left[(z_L - z_c) + (z_0 - z_L)\right]M_k(z_c).$$

Applying the second equation in Eq. (3.14) with $z_1 = z_L - z_c$ and $z_2 = z_0 - z_L$, we obtain the following *local expansion*:

$$\int_{S_c} G(z_0, z)q(z)dS(z) = \frac{1}{2\pi}\sum_{l=0}^{\infty} L_l(z_L)I_l(z_0 - z_L), \tag{3.24}$$

where the local expansion coefficients $L_l(z_L)$ are given by the following *moment-to-local (M2L) translation*:

$$L_l(z_L) = (-1)^l \sum_{k=0}^{\infty} O_{l+k}(z_L - z_c)M_k(z_c). \tag{3.25}$$

Similar to the multipole expansion, an estimate of the error bound for a local expansion with p terms from Eq. (3.24) can be found as follows [35]:

$$
\begin{aligned}
E_L^p &\equiv \left| \int_{S_c} G(z_0, z) q(z) dS(z) - \frac{1}{2\pi} \sum_{l=0}^{p} L_l(z_L) I_l(z_0 - z_L) \right| \\
&= \frac{1}{2\pi} \left| \sum_{l=p+1}^{\infty} L_l(z_L) I_l(z_0 - z_L) \right| \leq \frac{A \left[4e(p+\rho)(\rho+1) + \rho^2 \right]}{2\pi\rho(\rho-1)} \left(\frac{1}{\rho} \right)^{p+1}
\end{aligned}
$$

(3.26)

for any $p \geq \max\{2, \ 2\rho/(\rho - 1)\}$, where e is the base of the natural logarithm, and A and ρ are as defined for estimate (3.21).

It is interesting to note that we can also derive the preceding results in Eqs. (3.24) and (3.25) for the local expansion by starting from the following expression:

$$
G(z_0, z) = -\frac{1}{2\pi} \log(z_0 - z) = \frac{1}{2\pi} \sum_{k=0}^{\infty} O_k(z - z_L) I_k(z_0 - z_L), \quad (3.27)
$$

which is a Taylor series expansion of $G(z_0, z)$ about the point $z_0 = z_L$ that we can establish readily by using the Taylor series expansion as in Eq. (3.10). This expansion is symmetrical to the one in Eq. (3.11), which is an expansion of $G(z_0, z)$ about the point $z = z_c$.

We start with the expansion in Eq. (3.27) and evaluate:

$$
\begin{aligned}
&\int_{S_c} G(z_0, z) q(z) dS(z) \\
&= \frac{1}{2\pi} \sum_{k=0}^{\infty} \left[\int_{S_c} O_k(z - z_L) q(z) dS(z) \right] I_k(z_0 - z_L) \\
&= \frac{1}{2\pi} \sum_{k=0}^{\infty} \left[\int_{S_c} O_k \left((z - z_c) + (z_c - z_L) \right) q(z) dS(z) \right] I_k(z_0 - z_L) \\
&= \frac{1}{2\pi} \sum_{k=0}^{\infty} \left[\int_{S_c} \sum_{l=0}^{\infty} (-1)^l O_{k+l}(z_c - z_L) I_l(z - z_c) q(z) dS(z) \right] I_k(z_0 - z_L),
\end{aligned}
$$

where z_c is an expansion point near z with $|z - z_c| \ll |z_L - z_c|$ and the second relation in Eq. (3.14) has been applied. That is:

$$
\begin{aligned}
&\int_{S_c} G(z_0, z) q(z) dS(z) \\
&= \frac{1}{2\pi} \sum_{k=0}^{\infty} \left[\sum_{l=0}^{\infty} (-1)^l O_{k+l}(z_c - z_L) \int_{S_c} I_l(z - z_c) q(z) dS(z) \right] I_k(z_0 - z_L).
\end{aligned}
$$

(3.28)

Invoking the definition of the moment in Eq. (3.16), we can obtain Eqs. (3.24) and (3.25) for the local expansion from Eq. (3.28). This suggests that we can also establish the local expansion directly by simply defining the moment by using Eq. (3.16) without introducing the multipole expansion as given in Eq. (3.15).

3.2.5 Local-to-Local Translation

If the point for local expansion is moved from z_L to $z_{L'}$ (Figure 3.2), we have the following expression by using a local expansion with p terms from Eq. (3.24):

$$\int_{S_c} G(z_0, z)q(z)dS(z) \cong \frac{1}{2\pi}\sum_{l=0}^{p} L_l(z_L)I_l(z_0 - z_L)$$

$$= \frac{1}{2\pi}\sum_{l=0}^{p} L_l(z_L)I_l\left[(z_0 - z_{L'}) + (z_{L'} - z_L)\right].$$

Applying the first result in Eq. (3.14) and the relation $\sum_{l=0}^{p}\sum_{m=0}^{l} = \sum_{m=0}^{p}\sum_{l=m}^{p}$, we obtain:

$$\int_{S_c} G(z_0, z)q(z)dS(z) \cong \frac{1}{2\pi}\sum_{l=0}^{p} L_l(z_{L'})I_l(z_0 - z_{L'}), \qquad (3.29)$$

where the new coefficients are given by the following *local-to-local (L2L) translation*:

$$L_l(z_{L'}) = \sum_{m=l}^{p} I_{m-l}(z_{L'} - z_L)L_m(z_L). \qquad (3.30)$$

Replacing $m - l$ with m, we can also write (3.30) in an alternative form:

$$L_l(z_{L'}) = \sum_{m=0}^{p-l} I_m(z_{L'} - z_L)L_{l+m}(z_L). \qquad (3.31)$$

Note again that L2L translations involve only finite sums and do not introduce any new source of errors once the number of the local expansion terms p is fixed.

3.2.6 Expansions for the Integral with the *F* Kernel

We now consider the integral with the F kernel in CBIE (2.17) in complex notation:

$$\int_{S_c} F(z_0, z)\phi(z)dS(z), \qquad (3.32)$$

where $\phi(z)$ is still a real-valued function of complex variables z, and $F(z_0, z)$ is the F kernel in complex notation and can be written as:

$$F(z_0, z) = \frac{\partial G}{\partial n} = (n_1 + in_2)G' = n(z)G', \quad \text{with } G' \equiv \frac{\partial G}{\partial z}. \qquad (3.33)$$

Thus, the F kernel in real variables can be expressed as:

$$F(\mathbf{x}, \mathbf{y}) = \text{Re}\{F(z_0, z)\} = n_1 \text{ Re } G' - n_2 \text{ Im } G'. \tag{3.34}$$

From Eq. (3.11), we have:

$$G' = \frac{1}{2\pi} \sum_{k=1}^{\infty} O_k(z_0 - z_c) I_{k-1}(z - z_c), \tag{3.35}$$

and the integral in Eq. (3.32) becomes:

$$\int_{S_c} F(z_0, z)\phi(z)dS(z) = \frac{1}{2\pi} \sum_{k=1}^{\infty} O_k(z_0 - z_c)\widetilde{M}_k(z_c), \tag{3.36}$$

in which:

$$\widetilde{M}_k(z_c) = \int_{S_c} n(z) I_{k-1}(z - z_c)\phi(z)dS(z), \quad k = 1, 2, 3, \ldots, \tag{3.37}$$

are the *moments* for the F kernel integral, similar to those in Eq. (3.16) for the G kernel integral.

Again, on a constant element starting at point z_a and ending at z_b, we can evaluate the contribution to this moment analytically by using the relation in Eq. (3.18):

$$\widetilde{M}_k^{(e)}(z_c) = \int_{z_a}^{z_b} n(z) I_{k-1}(z - z_c)\phi(z)dS(z) = \phi_e n\overline{\omega} \left[I_k(z_b - z_c) - I_k(z_a - z_c) \right], \tag{3.38}$$

where ϕ_e is the nodal value of ϕ on this element.

The M2M, M2L, and L2L translations remain the same for the F kernel integral, except that $\widetilde{M}_0 = 0$. Therefore, all the translations used for M_k are applied for $\widetilde{M}_k$ directly.

3.2.7 Multipole Expansions for the Hypersingular Boundary Integral Equation

For the two integrals in HBIE (2.21), we can obtain the multipole expansions by directly taking derivatives of the related integrals in the CBIE. For example, for the K kernel integral, we have the following relations:

$$K(z_0, z) = \frac{\partial G}{\partial n(z_0)} = n(z_0)\frac{\partial G}{\partial z_0}, \tag{3.39}$$

$$\int_{S_c} K(z_0, z)q(z)dS(z) = n(z_0)\frac{\partial}{\partial z_0} \int_{S_c} G(z_0, z)q(z)dS(z)$$

$$= \frac{1}{2\pi}n(z_0) \sum_{l=1}^{\infty} L_l(z_L) I_{l-1}(z_0 - z_L),$$

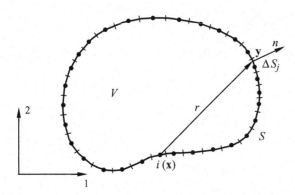

Figure 3.3. Discretization of the boundary S by use of constant elements.

by using Eq. (3.24) for the G kernel integral. That is, the *local expansion* for the K kernel integral in the HBIE is given by:

$$\int_{S_c} K(z_0, z)q(z)dS(z) = \frac{1}{2\pi}n(z_0)\sum_{l=0}^{\infty} L_{l+1}(z_L)I_l(z_0 - z_L), \qquad (3.40)$$

in which the same moments, M2M, M2L, and L2L translations for the G kernel integral in the CBIE can be applied directly. The same relation exists between the H kernel integral in the HBIE and the F kernel integral in the CBIE.

3.2.8 Fast Multipole Boundary Element Method Algorithms and Procedures

We are now ready to discuss the algorithms in the FMM for solving 2D potential problems by using the BEM. These fast multipole algorithms are the basic ones that can be extended readily to solve 3D potential problems and other 2D and 3D problems. Advanced algorithms, such as the adaptive algorithms, that can further speed up the solutions of the BEM equations also exist in the literature [60, 61].

An iterative solver, such as GMRES, is used to solve BEM equation (3.2). Each equation in this system of equations represents the sum of the integrals on all the elements when the source point is placed at one node. The FMM is used to evaluate the integrals on those elements that are far away from the source point, whereas the conventional approach is applied to evaluate the integrals on the remaining elements that are close to the source point. The detailed algorithms or procedures in the fast multipole BEM can be described as follows:

Step 1. *Discretization.* For a given problem, discretize the boundary S in the same way as in the conventional BEM approach. For example, we can apply constant elements to discretize the boundary S of a 2D domain, as shown in Figure 3.3.

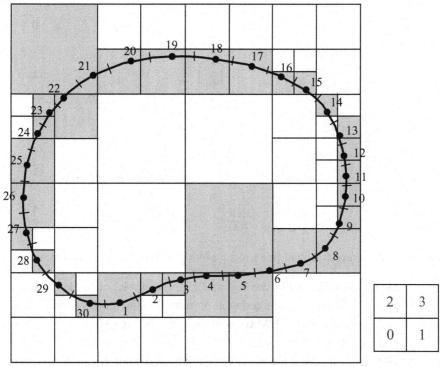

Figure 3.4. A hierarchical cell structure covering all of the boundary elements (the small square on the right-hand side shows the numbering scheme for the child cells of any given cell).

Step 2. *Determine a tree structure of the boundary element mesh.* For a 2D problem, we first consider a square that covers the entire boundary S and call this square the cell of level 0 (Figure 3.4). Then, we start dividing this *parent* cell into four equal *child* cells of level 1. Continue dividing in this way the cells that contain elements. For example, take a parent cell of level l and divide it into four child cells of level $l + 1$. Stop dividing a cell if the number of elements in that cell is fewer than a prespecified number (for illustration only, this number is taken as 1 in the example shown in Figure 3.4). A cell having no child cells is called a *leaf* (e.g., the shaded cells in Figure 3.4). Note that the edge length of a cell at level l is given by $L/2^l$, with L being the length of the edge of the largest cell at level 0. In this process, an element is considered to be within a cell if the center of the element is inside that cell. A *quad-tree* structure of the cells covering all the elements is thus formed after this procedure is completed (Figure 3.5).

Step 3. *Upward pass.* Compute the moments on all cells, at all levels with $l \geq$ 2, with up to p terms, and trace the tree structure upward (Figure 3.6). For a leaf, Eq. (3.16) is applied directly (with S_c being the set of the

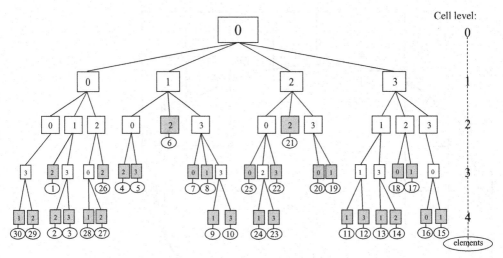

Figure 3.5. A hierarchical quad-tree structure for the 2D boundary element mesh.

elements contained in the leaf and z_c the centroid of the leaf). For a parent cell, calculate the moment by summing the moments on its four child cells using the M2M translation – that is, Eq. (3.23) – in which $z_{c\prime}$ is the centroid of the parent cell and z_c is the centroid of a child cell. Note that the moments need to be computed again for each new iteration of the solution because these moments involve the integration

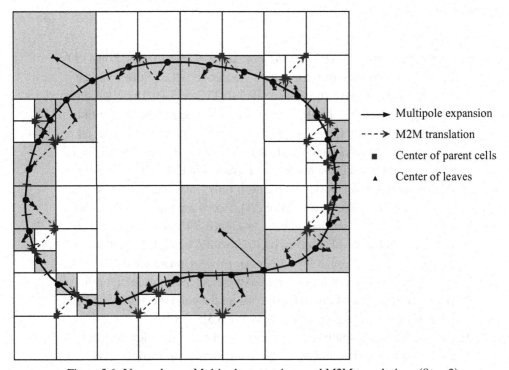

Figure 3.6. Upward pass: Multipole expansions and M2M translations (Step 3).

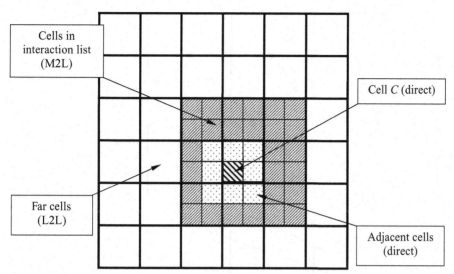

Figure 3.7. Grouping of the cells for cell *C* at level *l*.

of the kernels and estimated boundary solutions from the previous iteration.

Step 4. *Downward pass.* Let us first define a few terms used in describing the downward pass (Figure 3.7). Two cells are said to be *adjacent cells at level l* if they have at least one common vertex. (For two leaf cells at different levels, if the parent cell of one of the leaf cells shares at least a common vertex with the other leaf cell, they are also said to be adjacent cells.) Two cells are said to be *well separated at level l* if they are not adjacent at level *l* but their parent cells are adjacent at level *l* − 1. The list of all the well-separated cells from a level *l* cell *C* is called the *interaction list* of *C*. Cells are called to be *far cells* of *C* if their parent cells are not adjacent to the parent cell of *C*.

In the downward pass, we compute the local expansion coefficients on all cells starting from level 2 and tracing the tree structure downward to all the leaves (Figure 3.8). The local expansion associated with a cell *C* is the sum of the contributions from the cells in the interaction list of cell *C* and from all the far cells. The former is calculated by use of the M2L translation, Eq. (3.25), with moments associated with cells in the interaction list. The latter is calculated by use of the L2L translation, Eq. (3.30) or (3.31), for the parent cell of *C* with the expansion point being shifted from the centroid of *C*'s parent cell to that of *C*. For a cell *C* at level 2, we use only the M2L translation to compute the coefficients of the local expansion. Figure 3.8 shows how the local expansion coefficient is calculated through this downward pass for cell *C* where node 29 is located in our example model (see Figure 3.4).

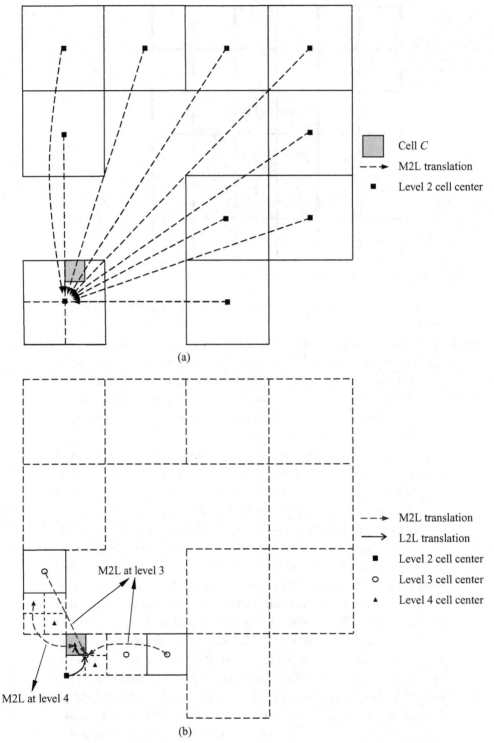

Figure 3.8. Downward pass: M2L and L2L translations (Step 4). (a) Level 2 cells; (b) levels 3 and 4 cells.

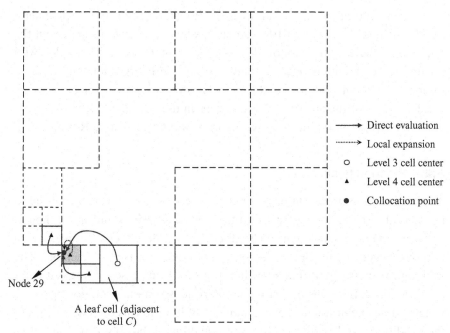

Figure 3.9. Evaluation of all the integrals for a collocation point (Step 5).

Step 5. *Evaluation of the integrals.* We use the G kernel integral in Eq. (3.8) as an example. Suppose the collocation point z_0 is on an element in leaf C (see Figure 3.7). We compute the contributions from elements in leaf C and its adjacent cells directly as in the conventional BEM.

We compute contributions from all other cells (cells in the interaction list of C and far cells) by using the local expansion; that is, Eq. (3.24). We do this by using the local expansion coefficients for cell C, which were computed in Step 4, and shifting the expansion point from the centroid of C to the collocation point z_0 (see Figure 3.6). That is, the integral is decomposed as follows:

$$\int_S G(z_0, z)q(z)dS(z) = \int_{S\_Near} GqdS + \int_{S\_Far} GqdS, \qquad (3.41)$$

where the integral on *S_Near* (cell C and its adjacent cells) is done by direct integration as in the conventional BEM, and the integral on *S_Far* (cells in the interaction list and far cells for cell C) is done by the FMM (M2L and L2L translations, respectively). Figure 3.9 shows how the evaluation of all the integrals is done for node 29 in our example model (see Figure 3.4).

Step 6. *Iterations of the solution.* The iterative solver updates the unknown solution vector λ in the system $\mathbf{A}\lambda = \mathbf{b}$ and continues at Step 3 to evaluate the next matrix and vector multiplication ($\mathbf{A}\lambda$) until the solution of λ converges within the given tolerance.

The fast multipole algorithm discussed in this section is the original algorithm, which is efficient for BEM models in which the elements are about the same size and distributed uniformly in a bulky domain. For BEM models with nonuniform element distributions and especially with large elements adjacent to smaller elements, the so-called *adaptive* FMMs are more efficient, in which the definitions of the adjacent cells and cells in the interaction list are further refined. Discussions on the adaptive algorithms can be found in Refs. [60, 61].

3.2.9 Preconditioning

Applying a good preconditioner for the iterative solver is very beneficial, if not crucial, for the convergence of the iterative solutions and the computational efficiency. Unlike that of the direct solver, the CPU time used by an iterative solver in solving a linear system of equations is unpredictable. The solution can converge within a few iterations for some cases, whereas it takes a few hundred iterations in other cases, depending on the conditioning of the system. It has been found that the number of iterations is directly related to the condition number of the system of equations to be solved with iterative solvers. To accelerate the iterative solution process – that is, to reduce the number of iterations for a given tolerance – a preconditioning matrix can be introduced to improve the conditioning of the BEM system matrix.

A simple and effective choice is to use a block diagonal preconditioner in the form:

$$\mathbf{M} = \begin{bmatrix} \mathbf{A}_1 & \mathbf{0} & \mathbf{0} & \cdots & \mathbf{0} \\ \mathbf{0} & \mathbf{A}_2 & \mathbf{0} & \cdots & \mathbf{0} \\ \mathbf{0} & \mathbf{0} & \mathbf{A}_3 & \cdots & \mathbf{0} \\ \vdots & \vdots & \vdots & \ddots & \mathbf{0} \\ \mathbf{0} & \mathbf{0} & \mathbf{0} & \mathbf{0} & \mathbf{A}_n \end{bmatrix}, \tag{3.42}$$

in which $\mathbf{A}_i$ is a submatrix of $\mathbf{A}$ with the coefficient formed on a leaf by direct evaluation of the integrals within that leaf. Using the preconditioner matrix $\mathbf{M}$, we change the original system:

$$\mathbf{A}\lambda = \mathbf{b} \tag{3.43}$$

to

$$\left(\mathbf{M}^{-1}\mathbf{A}\right)\lambda = \mathbf{M}^{-1}\mathbf{b} \tag{3.44}$$

for left preconditioning, or to:

$$\left(\mathbf{A}\mathbf{M}^{-1}\right)(\mathbf{M}\lambda) = \mathbf{b} \tag{3.45}$$

for right preconditioning, both of which can potentially yield better conditioned systems. Other forms of the preconditioners are also available, and it is still an important research topic to find a better preconditioner for the fast multipole BEM in many applications. Further discussion on the preconditioners for multidomain and elasticity problems is provided in the next chapter after the discussion of the fast multipole BEM for elasticity problems.

3.2.10 Estimate of the Computational Complexity

When the size of a BEM model is large, the estimated cost of the entire process just described for the fast multipole BEM is $O(N)$, with N being the number of elements or nodes, if the number of terms p in the multipole and local expansions and the maximum number of elements *maxl* allowed in a leaf are kept constant [41]. This claim on the $O(N)$ complexity of the fast multipole BEM is based on the following observations:

- N_{leaf} = number of leaves in the mesh $\cong N/maxl = O(N)$
- N_{cell} = number of cells $\cong N_{\text{leaf}} \times (1 + 1/4 + 1/4^2 + 1/4^3 + \cdots +) \leq N_{\text{leaf}} \times (4/3) = O(N)$
- Number of adjacent cells = 9; number of cells in the interaction list = 27 (for 2D models)
- Number of operations in computing multipole moments = $p \times maxl \times N_{\text{leaf}} = O(N)$
- Number of operations in upward pass = $N_{\text{cell}} \times 4 \times p^2 = O(N)$
- Number of operations in downward pass = $N_{\text{cell}} \times [p^2(\text{L2L}) + 27 \times p^2 (M2L)] = O(N)$
- Number of operations in local expansions = $N \times p = O(N)$
- Number of operations in direct evaluation of the integrals = $N \times 9 \times maxl = O(N)$

All the preceding estimates are at most $O(N)$; therefore, the total computational cost is also $O(N)$. These estimates will be slightly different for 3D cases and for dynamic problems. This $O(N)$ efficiency in computing for the fast multipole BEM is very significant when we solve large-scale problems, as will be demonstrated later through the numerical examples.

3.3 Programming for the Fast Multipole Boundary Element Method

We now discuss the main structure of a fast multipole BEM code for solving general 2D potential problems. This code, written in Fortran, is discussed in Ref. [62] and is provided in Appendix B.2. This fast multipole BEM code for general 2D potential problems can be used as the basis to develop fast multipole BEM programs for 3D potential, as well as 2D and 3D elasticity, Stokes flow, and acoustic wave problems, using constant or higher-order elements.

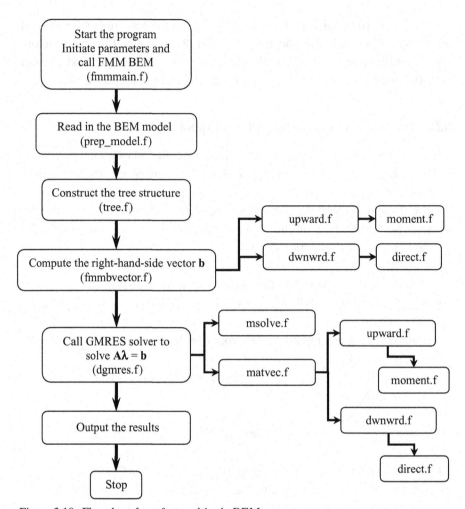

Figure 3.10. Flowchart for a fast multipole BEM program.

The flowchart of this fast multipole BEM code for the 2D potential code is given in Figure 3.10. The chart shows the main tasks for the program and the related subroutines (functions). The source code (dgmres.f) for the iterative solver GMRES (SLATEC GMRES package) can be downloaded from the *netlib* website (http://www.netlib.org/).

The program for the fast multipole BEM is much more involved than the program for the conventional BEM because of the tree structure of the cells and various expansions. Because of the restrictions of the SLATEC GMRES solver, a large array is needed in the program to pass the variables to the GMRES solver. Therefore, the main purpose of the main program is to allocate all the variables in this large array by calling the *lpointer* subroutine. Then, the subroutine for the fast multipole BEM, *fmmmain*, is invoked, which can be regarded as the starting point for the fast multipole BEM code. Explanations of all the main variables used in the program are given at the end of the main program (see Appendix B.2). A few important subroutines in the program are discussed in the following subsection, and other subroutines can be understood readily by reading the source code directly.

3.3.1 Subroutine *fmmmain*

The *fmmmain* subroutine starts with calling subroutine *prep_model*, which reads in the data for the boundary nodes, elements, boundary conditions, and field (interior) points from file *input.dat* (which is identical to the one used for the conventional BEM code in Appendix B.1), and the additional parameters used in the fast multipole expansions and solver GMRES from file *input.fmm* (a sample file is given in Appendix B.3). It then generates the tree structure, computes the right-hand-side $\mathbf{b}$ vector, solves the system of equations $\mathbf{A}\lambda = \mathbf{b}$ using the GMRES solver, computes values at interior points, and finally outputs the results.

3.3.2 Subroutine *tree*

We create the quad-tree structure for the elements by calling the subroutine *tree*, which is an essential piece of the entire code. The information of the tree structure is stored in several arrays in the code. To understand how this subroutine is used to create the tree structure, let us use the BEM model shown in Figure 3.4 as the example.

Cells in the tree structure are numbered in the following way: The largest cell at level 0 is called Cell 1, the four cells at level 1 are numbered 2, 3, 4, and 5, respectively, according to the order 0, 1, 2, 3, as shown in the side box in Figure 3.4. We continue in this way to level 2 cells and so on until we reach all the leaves. Empty cells (without any elements) are ignored. Cell numbers for the cells at levels 0, 1, and 2 for the model in Figure 3.4 are shown in Figure 3.11.

There are 30 elements in the model in Figure 3.4. The *tree* code sorts the elements in each cell (using the nodes, which are at the centers of the elements), first in the y direction and then in the x direction (twice) by dividing the elements into two groups according to the centerline in the related direction. This is done by invoking the subroutine *bisec*. Four child cells are formed after this process, which continues until a leaf is reached (in this example, each leaf contains only one element). The process can be illustrated as in Table 3.1, which produces a tree structure with 4 levels, 53 cells, and 30 leaves, as shown in Figure 3.5. The elements in the tree structure are rearranged (from left to right as shown in Table 3.1 and Figure 3.5); this information is stored in array *ielem*(k), which gives the original element number for the kth element in the tree structure.

Table 3.1. *Regrouping the elements using the* tree *code for the model in Figure 3.4*

Tree level	Sequences of the elements in the tree structure
0	1 2 3 4 5 6 7 8 9 10 11 12 13 14 15 16 17 18 19 20 21 22 23 24 25 26 27 28 29 30
1	1 2 3 26 27 28 29 30 4 5 6 7 8 9 10 19 20 21 22 23 24 25 11 12 13 14 15 16 17 18
2	29 30 1 2 3 26 27 28 4 5 6 7 8 9 10 22 23 24 25 21 19 20 11 12 13 14 17 18 15 16
3	29 30 1 2 3 27 28 26 4 5 6 7 8 9 10 25 23 24 22 21 20 19 11 12 13 14 18 17 15 16
4	30 29 1 2 3 28 27 26 4 5 6 7 8 9 10 25 24 23 22 21 20 19 11 12 13 14 18 17 16 15
ielem(*k*)	30 29 1 2 3 28 27 26 4 5 6 7 8 9 10 25 24 23 22 21 20 19 11 12 13 14 18 17 16 15

Five other arrays are used in the subroutine *itree* to store the additional information for the tree structure: *itree*, *loct*, *numt*, *ifath*, and *level*.

Array *itree*(*i*) gives the cell location of the *i*th cell within its corresponding tree level. At level *l*, the bounding square for the domain (Cell 0) is divided by $2^l \times 2^l$ grids. The numbering of the $2^l \times 2^l$ small squares starts from the lower-left corner with numbers 0, 1, 2, 3, …, and so on, first in the *x* direction, then in the *y* direction. For example, for Cell 5 in Figure 3.11, *itree*(5) = 3, and for Cell 12, *itree*(12) = 8. Values of *itree* can be used to determine the locations (coordinates) of the cells at all tree levels.

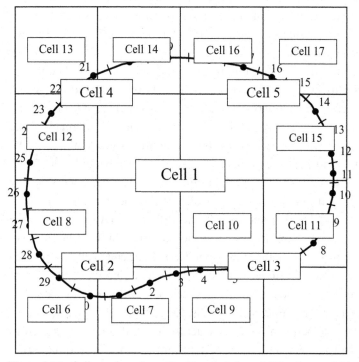

Figure 3.11. Cell numbers for cells at levels 0, 1, and 2 for the model in Figure 3.4.

Table 3.2. *Values of the arrays defining the tree structure for cells at levels 0, 1, and 2*

Cell no. i	$itree(i)$	$loct(i)$	$numt(i)$	$ifath(i)$
1	0	1	30	1
2	0	1	8	1
3	1	9	7	1
4	2	16	7	1
5	3	23	8	1
6	0	1	2	2
7	1	3	3	2
8	4	6	3	2
9	2	9	2	.3
10	3	11	1	3
11	7	12	4	3
12	8	16	4	4
13	12	20	1	4
14	13	21	2	4
15	11	23	4	5
16	14	27	2	5
17	15	29	2	5
⋮	⋮	⋮	⋮	⋮

Array $loct(i)$ indicates the starting place of the elements included in the ith cell in the array *ielem*. For example, for Cell 15 in Figure 3.11, $loct(15) = 23$.

Array $numt(i)$ gives the number of elements included in the ith cell. For example, $numt(3) = 7$ for Cell 3 in Figure 3.11.

Array $ifath(i)$ gives the cell number of the parent cell of the ith cell. For example, $ifath(1) = 0$ for Cell 1 and $ifath(11) = 3$ for Cell 11 in Figure 3.11.

The values of the arrays *itree*, *loct*, *numt*, and *ifath* for cells at levels 0, 1, and 2 for the model in Figure 3.11 are listed in Table 3.2, which we can use to understand the meanings of these arrays and the tree structure generated by the subroutine *tree*.

Finally, array $level(l)$ is used to indicate the starting cell number of all level l cells in the tree structure. For the model in Figure 3.11, $level(1) = 2$, $level(3) = 6$, and $level(3) = 18$.

3.3.3 Subroutine *fmmbvector*

After the tree structure is formed, we compute the right-hand-side **b** vector by using the fast multipole algorithms using the subroutine *fmmbvector*. We do this only once by calling the subroutines *upward* and *dwnwrd*. For large-scale models, using the FMM in computing the right-hand-side **b** vector can also save significant CPU time as compared with using the conventional direct method, which is also $O(N^2)$.

3.3.4 Subroutine *dgmres*

The *dgmres* subroutine is the GMRES solver in the SLATEC package from www.netlib.org. One does not need to understand the inner workings of this GMRES iterative solver to apply this subroutine. To use this GMRES solver, one needs to prepare only two subroutines: *msolve* and *matvec*, which are two external subroutines for *dgmres*.

The *msolve* subroutine prepares a preconditioning matrix for the iterative solver GMRES. In this program, the preconditioning matrix is formed by the block diagonal matrices based on the elements on the leaves. This preconditioning matrix is computed only once in the first iteration with the direct method and stored for use in all other iterations. This matrix is stored in the array *rwork*, and the related information (location and dimensions of each diagonal block matrix) is stored in array *iwork*.

The *matvec* subroutine provides the algorithm for the matrix–vector multiplication ($\mathbf{A}\lambda$) using the fast multipole algorithms by simply calling the *upward* and *dwnwrd* subroutines using the values for the solution vector from the previous iteration.

3.3.5 Subroutine *upward*

The *upward* subroutine calculates the multipole moments for all cells from leaves up to cells at level 2, climbing the tree structure upward and by using the boundary values from the previous iteration. For leaves, the moments are computed directly using the definition by calling the subroutine *moment*. For parent cells, M2M translations are applied to form the moments from the moments on their child cells.

3.3.6 Subroutine *dwnwrd*

The *dwnwrd* subroutine calculates the local expansions of the two integrals with G and F kernels at each source point. For far cells, the contributions are calculated with L2L translations. For cells in the interaction list, the contributions are calculated with M2L translations. For neighboring cells, direct integrations are applied by calling the *direct* subroutine, which is a variation of the *coefficient* subroutine used in the conventional BEM code (see Appendix B.1).

The 2D program just discussed can be extended readily to develop fast multipole BEM programs for 3D potentials and 2D or 3D elasticity, Stokes flow, and acoustic wave problems. For 3D problems, the major changes will be in the tree structure, in which the quad-tree structure for 2D problems is changed to an oct-tree structure.

3.4 Fast Multipole Formulation for 3D Potential Problems

In this section, the basic fast multipole expansions for 3D potential problems are discussed. We can implement the 3D fast multipole BEM by extending many of the results discussed in the previous sections for 2D potential problems. For example, the quad-tree structure for the elements in 2D problems is extended to an oct-tree structure of elements in 3D problems, in which the cells will be boxes and each parent cell will have eight child cells. The main structure of a computer program for 3D problems remains the same as the one discussed in the previous section for 2D problems.

First, we note that the kernel $G(\mathbf{x}, \mathbf{y})$ in Eq. (2.5) for 3D potential problems can be expanded as follows (see, e.g., Refs. [41, 61, 63]):

$$G(\mathbf{x}, \mathbf{y}) = \frac{1}{4\pi r} = \frac{1}{4\pi} \sum_{n=0}^{\infty} \sum_{m=-n}^{n} \overline{S_{n,m}}(\mathbf{x} - \mathbf{y}_c) R_{n,m}(\mathbf{y} - \mathbf{y}_c), \quad |\mathbf{y} - \mathbf{y}_c| < |\mathbf{x} - \mathbf{y}_c|,$$

(3.46)

where $\mathbf{y}_c$ is the expansion center close to the field point $\mathbf{y}$ and the overbar indicates the complex conjugate. The two functions $R_{n,m}$ and $S_{n,m}$ are called *solid harmonic* functions, given by:

$$R_{n,m}(\mathbf{x}) = \frac{1}{(n+m)!} P_n^m(\cos \theta) e^{im\phi} \rho^n,$$

(3.47)

$$S_{n,m}(\mathbf{x}) = (n-m)! P_n^m(\cos \theta) e^{im\phi} \frac{1}{\rho^{n+1}},$$

(3.48)

where (ρ, θ, ϕ) are the coordinates of $\mathbf{x}$ used here in a spherical coordinate system (specifically, $x_1 = \rho \sin \theta \cos \phi$, $x_2 = \rho \sin \theta \sin \phi$, $x_3 = \rho \cos \theta$) and P_n^m is the associated Legendre function. In this book, the following definition of the associated Legendre function is applied [64]:

$$P_n^m(x) = (1 - x^2)^{m/2} \frac{d^m}{dx^m} P_n(x),$$

(3.49)

where $P_n(x)$ is the Legendre polynomials of degree n [64]. In the literature, a slightly different definition exists for the associate Legendre function, in which a factor $(-1)^m$ is added to the right-hand side of Eq. (3.49).

The kernel $F(\mathbf{x}, \mathbf{y})$ for 3D potential problems can also be expanded as follows:

$$F(\mathbf{x}, \mathbf{y}) = \frac{\partial G(\mathbf{x}, \mathbf{y})}{\partial n(\mathbf{y})}$$

$$= \frac{1}{4\pi} \sum_{n=0}^{\infty} \sum_{m=-n}^{n} \overline{S_{n,m}}(\mathbf{x} - \mathbf{y}_c) \frac{\partial R_{n,m}(\mathbf{y} - \mathbf{y}_c)}{\partial n(\mathbf{y})}, \quad |\mathbf{y} - \mathbf{y}_c| < |\mathbf{x} - \mathbf{y}_c|.$$

(3.50)

Applying expansions in Eqs. (3.46) and (3.50), we can evaluate the G and F integrals in CBIE (2.17) on S_c (a subset of S that is away from source point $\mathbf{x}$) as follows:

$$\int_{S_c} G(\mathbf{x}, \mathbf{y}) q(\mathbf{y}) dS(\mathbf{y}) = \frac{1}{4\pi} \sum_{n=0}^{\infty} \sum_{m=-n}^{n} \overline{S_{n,m}}(\mathbf{x} - \mathbf{y}_c) M_{n,m}(\mathbf{y}_c), \quad |\mathbf{y} - \mathbf{y}_c| < |\mathbf{x} - \mathbf{y}_c|,$$

(3.51)

$$\int_{S_c} F(\mathbf{x}, \mathbf{y}) \phi(\mathbf{y}) dS(\mathbf{y}) = \frac{1}{4\pi} \sum_{n=0}^{\infty} \sum_{m=-n}^{n} \overline{S_{n,m}}(\mathbf{x} - \mathbf{y}_c) \tilde{M}_{n,m}(\mathbf{y}_c), \quad |\mathbf{y} - \mathbf{y}_c| < |\mathbf{x} - \mathbf{y}_c|,$$

(3.52)

where $M_{n,m}$ and $\tilde{M}_{n,m}$ are the *multipole moments* centered at $\mathbf{y}_c$ and defined as:

$$M_{n,m}(\mathbf{y}_c) = \int_{S_c} R_{n,m}(\mathbf{y} - \mathbf{y}_c) q(\mathbf{y}) dS(\mathbf{y}),$$

(3.53)

$$\tilde{M}_{n,m}(\mathbf{y}_c) = \int_{S_c} \frac{\partial R_{n,m}(\mathbf{y} - \mathbf{y}_c)}{\partial n(\mathbf{y})} \phi(\mathbf{y}) dS(\mathbf{y}).$$

(3.54)

When the multipole expansion center is moved from $\mathbf{y}_c$ to $\mathbf{y}_{c'}$, we apply the following M2M translation:

$$M_{n,m}(\mathbf{y}_{c'}) = \int_{S_c} R_{n,m}(\mathbf{y} - \mathbf{y}_{c'}) q(\mathbf{y}) dS(\mathbf{y}) = \sum_{n'=0}^{n} \sum_{m'=-n'}^{n'} R_{n',m'}(\mathbf{y}_c - \mathbf{y}_{c'}) M_{n-n',m-m'}(\mathbf{y}_c),$$

(3.55)

which is also valid for $\tilde{M}_{n,m}$.

The *local expansion* for the G kernel integral on S_c is given as:

$$\int_{S_c} G(\mathbf{x}, \mathbf{y}) q(\mathbf{y}) dS(\mathbf{y}) = \frac{1}{4\pi} \sum_{n=0}^{\infty} \sum_{m=-n}^{n} R_{n,m}(\mathbf{x} - \mathbf{x}_L) L_{n,m}(\mathbf{x}_L),$$

(3.56)

where the local expansion coefficients $L_{n,m}(\mathbf{x}_L)$ are given by the following M2L translation:

$$L_{n,m}(\mathbf{x}_L) = (-1)^n \sum_{n'=0}^{\infty} \sum_{m'=-n'}^{n'} \overline{S_{n+n',m+m'}}(\mathbf{x}_L - \mathbf{y}_c) M_{n',m'}(\mathbf{y}_c), \quad |\mathbf{x} - \mathbf{x}_L| < |\mathbf{y}_c - \mathbf{x}_L|,$$

(3.57)

in which $\mathbf{x}_L$ is the local expansion center.

The local expansion center can be shifted from $\mathbf{x}_L$ to $\mathbf{x}_{L'}$ by the following L2L translation:

$$L_{n,m}(\mathbf{x}_{L'}) = \sum_{n'=n}^{\infty} \sum_{m'=-n'}^{n'} R_{n'-n,m'-m}(\mathbf{x}_{L'} - \mathbf{x}_L) L_{n',m'}(\mathbf{x}_L). \qquad (3.58)$$

A similar local expansion and the same M2L and L2L translations are also valid for the F kernel integral with the moment $\widetilde{M}_{n,m}$.

For HBIE (2.21) in three dimensions, we can obtain the local expansions for the K and H integrals by taking the normal derivatives of the local expansions for the G and F integrals, respectively. For example, we have for the K kernel integral:

$$\int_{S_c} K(\mathbf{x},\mathbf{y}) q(\mathbf{y}) dS(\mathbf{y}) = \frac{1}{4\pi} \sum_{n=0}^{\infty} \sum_{m=-n}^{n} \frac{\partial R_{n,m}(\mathbf{x}-\mathbf{x}_L)}{\partial n(\mathbf{x})} L_{n,m}(\mathbf{x}_L), \qquad (3.59)$$

with $M_{n,m}$ in M2L translation (3.57). A similar local expansion exists for the H kernel integral in the HBIE. Therefore, the same moments, M2M, M2L, and L2L translations used for the G and F integrals in the CBIE can be used directly for the K and H integrals in the HBIE.

As mentioned previously, the implementation of the fast multipole BEM for 3D problems can be done readily by extending the results from the 2D case. First, the quad-tree structure used for 2D domains is replaced with an oct-tree structure, in which each cell in the oct-tree structure is a cube or a box. A parent cell will contain eight child cells for 3D problems. Other data structures are similar to those in the 2D case, and the 2D fast multipole BEM code discussed in the previous section can be modified readily to develop a code for solving 3D potential problems.

However, the fast multipole BEM for 3D problems is much more computing intensive than that for 2D problems because of the complexities of the expansions and translations required in the formulation. Careful considerations are needed in the computation of these expansions and translations; for example, using various recursive relations in evaluating the solid harmonic functions [63]. Adaptive algorithms [60, 61] based on further refined tree structures and a new version of the FMM using diagonal translations [61, 65, 66] have also been developed that can significantly improve the computational efficiencies for solving large-scale 3D potential problems.

An adaptive fast multipole BEM code for solving 3D potential problems based on the work in Ref. [61] can be found at the author's website (http://urbana.mie.uc.edu/yliu/Software), where the program and sample input files can be downloaded. This adaptive FMM BEM code is used in solving all the 3D examples in the following section.

Table 3.3. *Results of the potential and normal derivative for the annular region*

N	q_a Fast multipole BEM	q_a Conventional BEM	ϕ_b Fast multipole BEM	ϕ_b Conventional BEM
36	−401.7716	−401.7715	376.7237	376.7236
72	−400.4006	−400.4007	377.1410	377.1410
360	−400.0149	−400.0148	377.2548	377.2548
720	−400.0035	−400.0036	377.2579	377.2579
1440	−400.0007	−400.0005	377.2586	377.2586
2400	−400.0019	−400.0006	377.2588	377.2588
4800	−400.0016	−400.0006	377.2589	377.2589
7200	−399.9973	−399.9982	377.2588	377.2589
9600	−399.9977	−399.9969	377.2589	377.2589
Analytical solution	−400.0000		377.2589	

3.5 Numerical Examples

The same examples used in the previous chapter (see Section 2.12) with the conventional BEM approach are solved again with the fast multipole BEM programs. The accuracy and efficiency of the fast multipole BEM are compared with those of the conventional BEM.

3.5.1 An Annular Region

We first solve the same 2D potential problem as described in Section 2.12.1 and shown in Figure 2.11. For the fast multipole BEM, the numbers of terms for both moments and local expansions were set to 15, the maximum number of elements in a leaf to 20, and the tolerance for convergence of the solution to 10^{-8}. The fast multipole BEM results converged in 11 iterations for the smallest model (with 36 elements) and in 43 iterations for the largest model (with 9600 elements). These numbers can be reduced to 9 and 28 iterations, respectively, if the tolerance for convergence is increased to 10^{-6}.

Table 3.3 shows the results of ϕ_b and q_a obtained for this problem by use of the fast multipole BEM and compared with the conventional BEM as the total number of elements increases from 36 to 9600. As we can see, the fast multipole BEM is found to be as equally accurate as the conventional BEM with moderate values for the parameters in the fast multipole BEM. The CPU times used for both approaches in these calculations are plotted in Figure 3.12, which shows the significant advantage of the fast multipole BEM in savings compared with those of the conventional BEM. For example, for the largest model with 9600 elements, the fast multipole BEM used fewer than 17 s, whereas the conventional BEM used about 7500 s of CPU time on a laptop PC with a Pentium IV 2.4-GHz CPU.

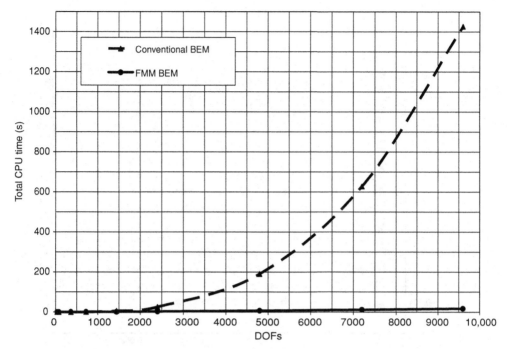

Figure 3.12. Comparison of the CPU times used by the conventional BEM and the FMM BEM.

3.5.2 Electrostatic Fields Outside Conducting Beams

We next study the simplified 2D models of comb drives used in microelectro-mechanical systems (MEMSs) by using the developed fast multipole BEM and comparing it with the conventional BEM. Both the CBIE and the dual BIE (CHBIE) formulations are used for this study. For the fast multipole BEM, the numbers of terms for both moments and local expansions are set to 15, the maximum number of elements in a leaf to 100, and the tolerance for convergence of the solutions to 10^{-6}.

The comb-drive models are built with the basic two-parallel-beam model used in Chapter 2 and shown in Figure 2.12. The parameters used are $\varepsilon = 1$, $L = 0.01$ m, $h = 0.0002$ m, $g = 0.0003$ m, $d = 0.0005$ m, and $V = 1$. Figure 3.13 shows a model with 17 beams. The two support beams on the left-hand and right-hand sides are not modeled in the BEM discretization. Two hundred elements are used along the beam length and five elements on each edge (with a total of elements equal to 410 for each beam). When more beams are added into the model, the number of elements along the beam length is increased to 400.

Figure 3.14 shows the computed charge densities on the center beam (beam 1) with positive voltage and the beam just below the center beam (beam 2) with negative voltage for the model with 17 beams shown in Figure 3.13.

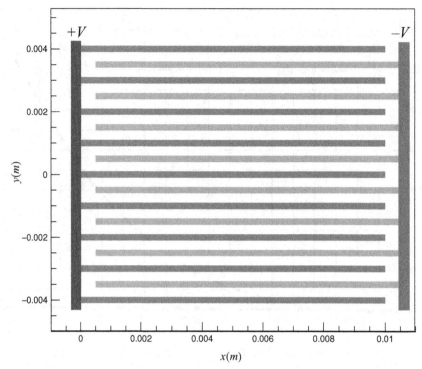

Figure 3.13. A 2D comb-drive model with 17 beams.

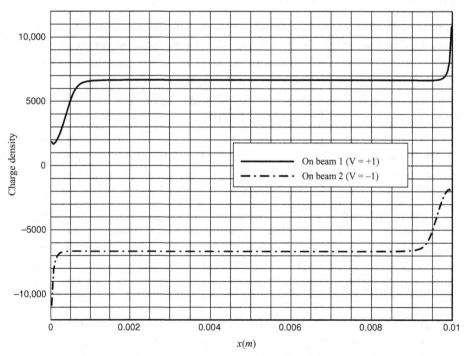

Figure 3.14. Charge densities on center beam 1 and beam 2 (below the center beam).

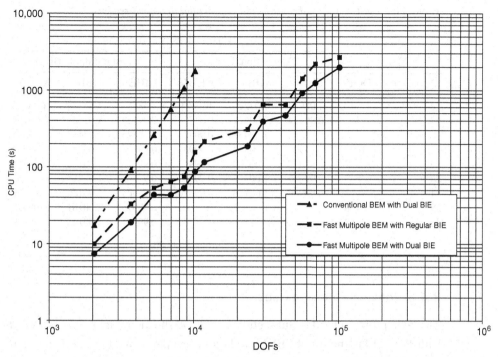

Figure 3.15. CPU times for the conventional BEM and fast multipole BEM.

Because of the symmetry of the fields above and below each beam, the charge densities on the top and bottom surfaces of each beam are identical; thus, only one field is plotted for each beam. The charge densities on the two beams are also of opposite sign and "antisymmetrical," as expected. It should be noted that the fields in MEMS are more complicated than those that the simple parallel-beam models can represent, especially near the edges of the beams, because of the simplified geometries used.

Figure 3.15 shows the CPU time comparison in which the conventional BEM and the fast multipole BEM are used in solving these simple comb-drive models on the 2.4-GHz Pentium IV laptop PC. Again, the conventional BEM can solve models only with up to 10,000 DOFs. Conversely, the fast multipole BEM with the dual BIE converges faster than the one with the regular BIE (CBIE alone) because of the better conditioning of the dual BIE formulation. The fast multipole BEM results converge in about 30 to 70 iterations when the dual BIE is used and in about 50 to more than 100 iterations when the regular BIE is used. It is evident from these studies that the dual BIE is very effective in solving MEMS problems with thin beams and the fast multipole BEM using the dual BIE is very efficient in solving large-scale 2D models.

Table 3.4. *Results for the cube with a linear potential in the x direction*

| Model | | Charge density at $(0.5, 0, 0)$ | | | | | |
| Elem/edge | DOFs | Conventional BEM | | | Fast multipole BEM | | |
		CBIE	HBIE	CHBIE	CBIE	HBIE	CHBIE
2	48	1.08953	1.07225	1.06800	1.08955	1.07278	1.06843
4	192	0.99124	1.00624	0.99754	0.99124	1.00624	0.99754
8	768	0.99825	1.00438	0.99894	0.99825	1.00438	0.99894
12	1728	0.99908	1.00327	0.99934	0.99908	1.00327	0.99934
16	3072	0.99942	1.00260	0.99953	0.99943	1.00260	0.99953
20	4800	0.99959	1.00216	0.99963	0.99962	1.00218	0.99965
24	6912	0.99969	1.00185	0.99970	0.99969	1.00184	0.99969
28	9408	–	–	–	0.99976	1.00161	0.99975
32	12288				0.99981	1.00143	0.99979
Exact value				1.00000			

3.5.3 Potential Field in a Cube

The cube problem used in Subsection 2.12.3 and shown in Figure 2.15 is solved with the 3D fast multipole BEM code and compared with the conventional BEM. For the fast multipole BEM, 15 terms are used in all of the expansions and the tolerance for convergence is set to 10^{-6}.

Table 3.4 shows the results obtained with the fast multipole BEM and compared with those of the conventional BEM, using the CBIE, HBIE, and CHBIE, for BEM meshes with increasing numbers of elements. We can conclude from these results that the HBIE and CHBIE are equally as accurate as the CBIE; so is the fast multipole BEM compared with the conventional BEM. Constant triangular elements are used in this study. If linear or quadratic elements were applied, a few elements should have been sufficient for obtaining results of a similar accuracy because of the specified linear field.

3.5.4 Electrostatic Field Outside Multiple Conducting Spheres

In this example, 11 perfectly conducting spheres (Figure 3.16) are analyzed with the fast multipole BEM. The center large sphere has a radius of 3; the 10 small spheres have the same radius of 1, and are distributed evenly on a circle with a radius of 5 and cocentered with the large sphere. A constant electric potential $\phi = +5$ is applied to the large sphere and five of the small spheres, and a potential $\phi = -5$ is applied to the other five small spheres (Figure 3.16). For the fast multipole BEM, elements per leaf are limited to 200, 10 terms are used in the expansions, and the tolerance for convergence is set to 10^{-4}.

The charge densities on the surfaces of the spheres are plotted in Figure 3.17 with the mesh using 10,800 elements per sphere. The plots are almost

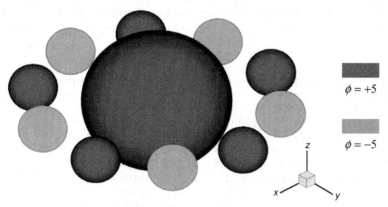

Figure 3.16. An 11-spherical perfect conductor model.

identical among the different meshes and exhibit the same symmetrical pattern, as it should be. Table 3.5 shows the maximum and minimum values of the charge densities on the spheres when the different meshes are used. These values are very stable and converged within the first two significant digits (except for the last set of data with the CBIE). Further improvements can be achieved by using a tighter set of parameters for the fast multipole BEM (e.g., more expansion terms and smaller tolerance). The last two columns of Table 3.5 show the numbers of iterations with the GMRES solver for the CBIE and the CHBIE. The numbers of iterations for the CHBIE is about half those for the CBIE because of the better conditioning of the systems of equations based on the CHBIE.

3.5.5 A Fuel Cell Model

Next, an example of more challenging problems is presented. Figure 3.18(a) shows a solid oxide fuel cell (SOFC) model with nine cells used for thermal analysis. There are 1000 small holes on the inner and outer surfaces of each

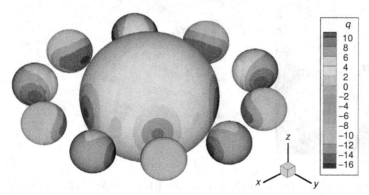

Figure 3.17. Contour plot of the charge densities on the spheres.

Table 3.5. *Results for the 11-sphere model obtained with the fast multipole BEM*

Model		Charge densities on the spheres				Numbers of iterations	
		min		max			
Elem/sphere	DOFs	CBIE	CHBIE	CBIE	CHBIE	CBIE	CHBIE
768	8448	−16.4905	−15.5285	11.1837	10.3923	14	8
1200	13200	−16.5363	−15.7922	11.2218	10.5920	15	8
1728	19008	−16.6322	−15.9618	11.2558	10.7156	17	8
2352	25872	−16.6436	−16.0789	11.2746	10.8041	18	8
3072	33792	−16.6733	−16.1618	11.3792	10.9160	19	8
3888	42768	−16.6648	−16.2195	11.3810	10.9464	20	7
4800	52800	−16.7435	−16.2671	11.3787	10.9763	20	8
7500	82500	−16.7068	−16.3614	11.2964	11.0283	21	8
10800	118800	−17.1157	−16.4279	12.6511	11.0851	22	7

cylindrical cell, with a total of 9000 holes for the entire stack model. Because of the extremely complicated geometry, the FEM (e.g., ANSYS®) can model only one cell on a PC with 1-GB RAM. For the fast multipole BEM, however, multicell models can be handled readily, such as the nine-cell stack modeled successfully with 530,230 elements and solved on a desktop PC with 1-GB RAM [Figure 3.18 (b)].

3.5.6 Image-Based Boundary Element Method Models and Analysis

In recent years, digital models using 3D scanning technologies have attracted much attention in many engineering fields, such as reverse engineering and biomedical engineering applications. Computer-scanned images are often

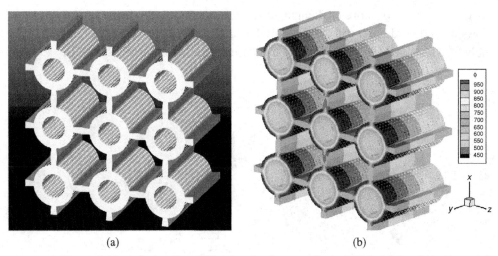

(a)   (b)

Figure 3.18. A fuel cell model using the fast multipole BEM: (a) 3 × 3 stack model; (b) computed temperature.

complicated in geometry and difficult to mesh and analyze with the domain-based methods because of the lack of the volume data from the scanned images. The scanned data are surface-based and usually in stereolithography (STL), nonuniform rational B spline (NURBS), and other file formats. Construction of the volume using these surface data is time-consuming and often inaccurate. Conversely, meshing the boundary of a scanned object using the surface data is straightforward and can be as accurate as the resolution of the scanner allows.

The fast multipole BEM seems to be a very natural choice to be integrated for the image-based analysis of various engineering problems. Boundary meshes can be obtained quickly from the scanned surface data, especially data in the STL format. Fast and accurate analysis using the fast multipole BEM can then be obtained. The potentials of the integration of the fast multipole BEM with 3D imaging technologies are huge in applications of reverse engineering, material characterizations, and biomedical applications.

A couple of examples are presented here to show the potential of the image-based analysis with the fast multipole BEM. This work is described in more detail in Ref. [67]. Figure 3.19 shows oil-lamp models generated by a 3D laser scanner and analyzed by both the FEM (ANSYS® software) and the 3D fast multipole BEM code [61]. The FEM volume mesh contains 403,271 tetrahedral elements, whereas the BEM mesh has 42,810 triangular elements to maintain a similar surface mesh density as in the FEM mesh. The top of the lamp is applied with a temperature of one unit and the bottom with a zero temperature. The other surfaces have zero-flux BCs. The two computed temperature results are comparable, as shown in the figure. The CPU times are close to 1 h for the ANSYS solution and less than 15 mins for the fast multipole BEM simulation, computed on a 3.2-GHz Pentium IV desktop PC.

Figure 3.20 shows a microscale model and thermal analysis of a weak trabecular bone sample using a 3D microscanner together with the fast multipole BEM code. There are about 200,000 elements in this model, and the model was solved in 3.4 h on the Pentium IV PC. The longer CPU time in solving this model is due to the increased number of iterations. Because of the many thin shapes in this complicated model, the conditioning of the BEM system of equations worsened; thus, it requires more iterations when the iterative solver is used. More discussions of the preceding results can be found in Ref. [67].

All of the preceding numerical examples clearly demonstrate the accuracy and efficiency of the fast multipole BEM for solving large-scale 2D and 3D potential problems. In all of the cases, constant elements were applied to implement the fast multipole BEM. Constant elements can certainly be replaced with higher-order elements to improve the accuracy of a fast multipole BEM code. However, this may not be advantageous, considering the

Figure 3.19. Image-based thermal analysis of an oil lamp model: (a) FEM volume mesh, (b) FEM temperature results, (c) BEM surface mesh, (d) BEM temperature results.

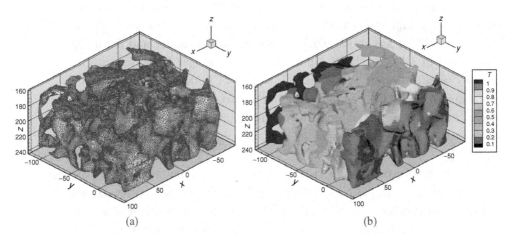

Figure 3.20. Image-based thermal analysis of a trabecular bone microstructure: (a) BEM surface mesh, (b) BEM temperature results.

efficiency of the code for large-scale problems. For constant elements, all the integrals can be evaluated analytically for all nonsingular, nearly singular, and singular cases. There are no numerical integrations in the code. Therefore, the code can be very efficient. For higher-order elements, however, this is not the case, and we have to use numerical integration in the direct evaluations of the integrals that can involve singular and nearly singular integrals. This complicates the code and reduces the efficiency of the fast multipole BEM solutions for large-scale problems.

3.6 Summary

An introduction of the fast multipole BEM is presented in this chapter for 2D and 3D potential problems. The main idea of the fast multipole BEM is to replace the element-to-element interactions, which are costly to compute, with cell-to-cell interactions through the introduction of the multipole expansions of the kernels and related translations that are integrated with a hierarchical tree structure of the boundary elements. Complete formulations and implementation details of the fast multipole BEM are provided in this chapter. The Fortran code provided in Appendix B is also discussed; it can be used to solve 2D potential problems, to learn the structure of a fast multipole BEM code, and to expand it to solve other large-scale 2D and 3D problems. Several numerical examples are presented to demonstrate the accuracy, efficiencies, and usefulness of the fast multipole BEM for solving large-scale 2D and 3D potential problems, especially in new technologies such as image-based modeling and simulations in reverse engineering and biomedical engineering. The fast multipole BEM algorithms and code presented in this chapter are the essence of the discussions in this book that should be studied thoroughly before one embarks on studying other topics in the fast multipole BEM. The approaches and the code discussed in this chapter can also be extended readily to solve 2D and 3D vector (elastostatic and Stokes flow) and 2D and 3D acoustic problems, as well as many other problems in applied mechanics.

Problems

3.1. Show that for two functions $f_1(N) = aN^2$ and $f_2(N) = bN$, one can always have $f_1 \gg f_2$ for sufficiently large N, no matter how small the value of a and how large the value of b can be.

3.2. Verify Eq. (3.6) with Eq. (3.7); that is, the real part of $G(z_0, z)$ in (3.7) does give the real-valued Green's function $G(\mathbf{x}, \mathbf{y})$ in real variables.

3.3. Derive expression (3.11) for the complex Green's function by using a Taylor series expansion.

3.4. Derive the L2L translation given in Eq. (3.30).

3.5. Verify Eq. (3.34) with Eq. (3.33).

3.6. Write a computer code to generate the quad-tree structure shown in Figure 3.5 for the boundary element mesh shown in Figure 3.4.

3.7. Continuing Problem 2.9 in Chapter 2, apply the 2D fast multipole BEM code to the quarter-symmetry annular region model. Compare the accuracy and efficiency of the results obtained with the conventional BEM and the fast multipole BEM.

3.8. Develop a 2D potential fast multipole BEM code using *linear* elements, based on the 2D potential fast multipole BEM code given in Appendix B.2 and with constant elements. Compare the accuracy and computational efficiency of the developed code with those of the code using constant elements.

3.9. Develop a 3D potential fast multipole BEM code using *constant triangular* elements by extending the 2D potential fast multipole BEM code given in Appendix B.2.

4 Elastostatic Problems

The direct BIE formulation and its numerical solutions using the BEM for 2D elasticity problems were developed by Rizzo in the early 1960s and published in Ref. [4] in 1967. Following this early work, extensive research efforts were made for the development of the BIE and BEM for solving various elasticity problems (see, e.g., Refs. [24–28]). The advantages of the BEM for solving elasticity problems are the accuracy in modeling stress concentration or fracture mechanics problems and the ease in modeling complicated elastic domains such as various composite materials.

The FMM was applied to solving elasticity problems for more than a decade. For 2D elasticity problems, Greengard et al. [68, 69] developed a fast multipole formulation for solving the biharmonic equations using potential functions. Peirce and Napier [36] developed a spectral multipole approach that shares some common features with the FMMs. Richardson et al. [70] proposed a similar spectral method using both 2D conventional and traction BIEs in the regularized form. Fukui [71] and Fukui et al. [72] studied both the conventional BIE for 2D stress analysis and the HBIE for large-scale crack problems. In his work, he first applied the complex variable representation of the kernels and then used the multipole expansions in complex variables as originally used for 2D potential problems [35, 62]. Liu [73, 74] further improved Fukui's approach and proposed a new set of moments for 2D elasticity CBIEs, which yields a very compact and efficient formulation with all the translations being symmetrical regarding the two sets of moments. Wang and Yao [75] also studied crack problems by using a dual BIE approach, with the CBIE collocating on one surface of a crack and HBIE on the other. They expanded the kernel functions in their original forms by using complex Taylor series in an auxiliary way following the approach in Ref. [76].

For 3D elasticity problems, Fu et al. [38] formulated the BIE for 3D elastic inclusion problems by using the FMM. Some other earlier development of the fast multipole BEM for general 3D elasticity problems can be found in Ref. [77] and for crack problems in Refs. [39, 78, 79]. Large-scale modeling of composite materials using the fast multipole elasticity BEM can be found in Refs. [80–82].

In this chapter, the governing equations for elasticity problems are reviewed first. Then, the fundamental solutions are introduced and the BIEs are established. The conventional BEM approach is discussed briefly, followed by discussions on the FMM for solving the BIEs for 2D and 3D elasticity problems in both single and multiple domains. Numerical examples are provided to demonstrate the accuracy and efficiencies of the fast multipole BEM for solving large-scale elasticity problems.

4.1 The Boundary-Value Problem

Consider the displacement u_i, strain ε_{ij}, and stress σ_{ij} in a linearly elastic solid occupying domain V with boundary S. The governing equations for these elastic fields are as follows:

Equilibrium equations:

$$\sigma_{ij,j} + f_i = 0, \quad \text{in } V, \tag{4.1}$$

where f_i is the body force.

Strain–displacement relation:

$$\varepsilon_{ij} = \frac{1}{2}(u_{i,j} + u_{j,i}), \quad \text{in } V. \tag{4.2}$$

Stress–strain relation (constitutive equations):

$$\sigma_{ij} = E_{ijkl}\varepsilon_{kl}, \quad \text{in } V, \tag{4.3}$$

where E_{ijkl} is the elastic modulus tensor given by:

$$E_{ijkl} = \lambda\delta_{ij}\delta_{kl} + \mu(\delta_{ik}\delta_{jl} + \delta_{il}\delta_{jk}) \tag{4.4}$$

for isotropic materials, and λ and μ are the Lamé constants that are related to Young's modulus E and Poisson's ratio v by:

$$\lambda = \frac{Ev}{(1+v)(1-2v)}, \quad \mu = \frac{E}{2(1+v)}. \tag{4.5}$$

The boundary conditions for an elasticity problem can be described by:

$$u_i = \overline{u_i} \quad \text{on } S_u \text{ (displacement BC)}, \tag{4.6}$$

$$t_i = \sigma_{ij}n_j = \overline{t_i} \quad \text{on } S_t \text{ (traction BC)}, \tag{4.7}$$

where the overbar indicates the given value, t_i is the traction, n_i are the components of the outward normal, and $S_u \cup S_t = S$.

The main objective in elasticity is to solve for the fields u_i, ε_{ij}, and σ_{ij} using governing equations (4.1), (4.2), and (4.3) under the BCs in (4.6) and (4.7).

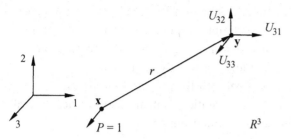

Figure 4.1. An infinite elastic domain applied with a unit concentrate force P at $\mathbf{x}$.

4.2 Fundamental Solution for Elastostatic Problems

Consider the full infinite space (R^2 for two dimensions or R^3 for three dimensions) filled by an elastic material. Apply a unit concentrate force P at point $\mathbf{x}$ in the ith direction. The responses (displacement, strain, and stress) at any point $\mathbf{y}$ that are due to this unit force are called the *fundamental solution* (or *Kelvin's solution*) in elasticity (Figure 4.1).

The stress component $\Sigma_{ijk}(\mathbf{x}, \mathbf{y})$ in the fundamental solution satisfies the following equilibrium equation:

$$\Sigma_{ijk,k}(\mathbf{x}, \mathbf{y}) + \delta_{ij}\delta(\mathbf{x}, \mathbf{y}) = 0, \quad \forall \mathbf{x}, \mathbf{y} \in R^2/R^3, \tag{4.8}$$

where $(\,)_{,k} = \partial (\,)/\partial y_k$, the first index i indicates the direction of the unit concentrated force at the source point $\mathbf{x}$, and the Dirac δ function $\delta(\mathbf{x}, \mathbf{y})$ represents the body force corresponding to the unit concentrated force.

For 2D (plane-strain) problems, the displacement and traction components in the fundamental solution are given by:

$$U_{ij}(\mathbf{x}, \mathbf{y}) = \frac{1}{8\pi\mu(1-v)}\left[(3-4v)\delta_{ij}\log\left(\frac{1}{r}\right) + r_{,i}\,r_{,j} - \frac{1}{2}\delta_{ij}\right], \tag{4.9}$$

$$T_{ij}(\mathbf{x}, \mathbf{y}) = -\frac{1}{4\pi(1-v)r}\left\{\frac{\partial r}{\partial n}\left[(1-2v)\delta_{ij} + 2r_{,i}\,r_{,j}\right] - (1-2v)\left(r_{,i}\,n_j - r_{,j}\,n_i\right)\right\}, \tag{4.10}$$

in which the index i indicates the direction of the unit force at the source point $\mathbf{x}$ and the index j indicates the jth component of the field at the field point $\mathbf{y}$. For plane-stress problems, Poisson's ratio v in the preceding expressions is replaced with $v/(1+v)$.

For 3D problems, the fundamental solution gives:

$$U_{ij}(\mathbf{x}, \mathbf{y}) = \frac{1}{16\pi\mu(1-v)r}\left[(3-4v)\delta_{ij} + r_{,i}\,r_{,j}\right], \tag{4.11}$$

$$T_{ij}(\mathbf{x}, \mathbf{y}) = -\frac{1}{8\pi(1-v)r^2}\left\{\frac{\partial r}{\partial n}\left[(1-2v)\delta_{ij} + 3r_{,i}\,r_{,j}\right] - (1-2v)\left(r_{,i}\,n_j - r_{,j}\,n_i\right)\right\}. \tag{4.12}$$

It is interesting to note that the fundamental solution for elasticity problems is closely related to the fundamental solution for potential problems. Both fundamental solutions have the same order of singularities as their corresponding 2D and 3D counterparts. For example, U_{ij} is weakly singular and T_{ij} is strongly singular, similar to G and F, respectively, for the potential problems.

The fundamental solution for elastostatic problems also satisfies several *integral identities* [46–48] as given in the following equations:

First identity:

$$\int_S T_{ij}(\mathbf{x}, \mathbf{y}) dS(\mathbf{y}) = \begin{cases} -\delta_{ij}, & \forall \mathbf{x} \in V \\ 0, & \forall \mathbf{x} \in E \end{cases}. \tag{4.13}$$

Second identity:

$$\int_S \frac{\partial T_{ij}(\mathbf{x}, \mathbf{y})}{\partial x_k} dS(\mathbf{y}) = 0, \quad \forall \mathbf{x} \in V \cup E. \tag{4.14}$$

Third identity:

$$E_{jlpq} \int_S \frac{\partial U_{iq}(\mathbf{x}, \mathbf{y})}{\partial x_k} n_p(\mathbf{y}) dS(\mathbf{y}) - \int_S \frac{\partial T_{ij}(\mathbf{x}, \mathbf{y})}{\partial x_k} (y_l - x_l) dS(\mathbf{y}) = \begin{cases} \delta_{ij}\delta_{kl}, & \forall \mathbf{x} \in V \\ 0, & \forall \mathbf{x} \in E \end{cases}. \tag{4.15}$$

Fourth identity:

$$\int_S T_{ij}(\mathbf{x}, \mathbf{y})(y_k - x_k) dS(\mathbf{y}) - E_{jkpq} \int_S U_{ip}(\mathbf{x}, \mathbf{y}) n_q(\mathbf{y}) dS(\mathbf{y}) = 0, \quad \forall \mathbf{x} \in V \cup E, \tag{4.16}$$

where S is an arbitrary and *closed* contour (for two dimensions) or surface (for three dimensions), V is the domain enclosed by S, and E is the infinite (exterior) domain outside S. These identities have clear physical meanings and can be very convenient in deriving various weakly singular forms of the BIEs for elasticity problems [46–48]. These identities can be derived readily by integrating governing equation (4.8) over the domain V and invoking the Gauss theorem [46–48].

4.3 Boundary Integral Equation Formulations

To derive the BIEs for elastostatic problems, we first establish the generalized Green's identity corresponding to elasticity equations. Let $(u_i, \varepsilon_{ij}, \sigma_{ij})$ and $(u_i^*, \varepsilon_{ij}^*, \sigma_{ij}^*)$ be two sets of solutions satisfying governing equations (4.1)–(4.3) in domain V. The following generalized Green's identity, also called *Somigliana's identity*, holds:

$$\int_V \left(\sigma_{jk,k} u_j^* - \sigma_{jk,k}^* u_j \right) dV = \int_S \left(t_j u_j^* - t_j^* u_j \right) dS. \tag{4.17}$$

This identity can be derived readily by using either the Gauss theorem or the virtual work theorem.

Now, let $(u_i, \varepsilon_{ij}, \sigma_{ij})$ be the solution of the boundary-value problem that needs to be solved, and let $(u_i^*, \varepsilon_{ij}^*, \sigma_{ij}^*)$ be the fundamental solution; that is:

$$u_j^*(\mathbf{y}) = U_{ij}(\mathbf{x}, \mathbf{y}), \quad t_j^*(\mathbf{y}) = T_{ij}(\mathbf{x}, \mathbf{y}), \quad \sigma_{jk,k}^*(\mathbf{y}) = \Sigma_{ijk,k}(\mathbf{x}, \mathbf{y}).$$

Substituting these results into identity (4.17) and applying Eqs. (4.1) and (4.8), we obtain the following *representation integral* of the displacement field in domain V:

$$u_i(\mathbf{x}) = \int_S [U_{ij}(\mathbf{x}, \mathbf{y})t_j(\mathbf{y}) - T_{ij}(\mathbf{x}, \mathbf{y})u_j(\mathbf{y})]\, dS(\mathbf{y})$$

$$+ \int_V U_{ij}(\mathbf{x}, \mathbf{y}) f_j(\mathbf{y}) dV(\mathbf{y}), \quad \forall \mathbf{x} \in V. \tag{4.18}$$

Once the displacement u_i and traction t_i are obtained on the entire boundary S, the preceding expression can be used to evaluate the displacement at any point inside the domain V, if needed.

Let the source point $\mathbf{x}$ approach boundary S in Eq. (4.18) in the same way as discussed in Chapter 2 for the BIE for potential problems; we obtain the following conventional BIE (CBIE) for elastostatic problems:

$$c_{ij}(\mathbf{x})u_j(\mathbf{x}) = \int_S [U_{ij}(\mathbf{x}, \mathbf{y})t_j(\mathbf{y}) - T_{ij}(\mathbf{x}, \mathbf{y})u_j(\mathbf{y})]\, dS(\mathbf{y})$$

$$+ \int_V U_{ij}(\mathbf{x}, \mathbf{y}) f_j(\mathbf{y}) dV(\mathbf{y}), \quad \forall \mathbf{x} \in S, \tag{4.19}$$

where the coefficients $c_{ij} = \frac{1}{2}\delta_{ij}$ if S is smooth at source point $\mathbf{x}$. In general, we have the following expression for c_{ij}:

$$c_{ij}(\mathbf{x}) = \delta_{ij} + \lim_{\varepsilon \to 0} \int_{S_\varepsilon(\mathbf{x})} T_{ij}(\mathbf{x}, \mathbf{y}) dS(\mathbf{y}) = \gamma \delta_{ij} - \int_S T_{ij}(\mathbf{x}, \mathbf{y}) dS(\mathbf{y}), \tag{4.20}$$

in which $\gamma = 0$ for finite domain problems and $\gamma = 1$ for infinite domain problems and the last integral is a CPV integral. In deriving the preceding result, the first identity in Eq. (4.13) is applied.

In CBIE (4.19), the integral with the U kernel is a weakly singular integral, whereas the integral with the T kernel is a strongly singular (CPV) integral. CBIE (4.19) can be applied to solve for the unknown displacement and traction on the boundary.

The domain integral in CBIE (4.19) can be handled with the approaches presented in Section 2.9 in the case in which $f_j(\mathbf{y})$ is nonzero over a finite area or volume within the domain V. If $f_j(\mathbf{y})$ is due to *a concentrated or point force* within V, we can write $f_j(\mathbf{y})$ as:

$$f_j(\mathbf{y}) = Q_j \delta(\mathbf{x}_Q, \mathbf{y}), \tag{4.21}$$

where $\mathbf{x}_Q$ is the location of the concentrated force and Q_j represents the components of the concentrated force. Using the sifting property of the Dirac δ

function [Eq. (1.25)], we can evaluate the domain integral in CBIE (4.19) for a concentrated force readily as follows:

$$\int_V U_{ij}(\mathbf{x}, \mathbf{y}) f_j(\mathbf{y}) dV(\mathbf{y}) = Q_j \int_V U_{ij}(\mathbf{x}, \mathbf{y}) \delta(\mathbf{x}_Q, \mathbf{y}) dV(\mathbf{y}) = Q_j U_{ij}(\mathbf{x}, \mathbf{x}_Q).$$
$$(4.22)$$

This contribution is added to the right-hand-side vector **b** of the BEM system of equations based on CBIE (4.19).

Taking the derivatives of representation integral (4.18), applying the stress–strain relation, and letting the source point **x** go to the boundary, we can obtain the traction or HBIE as follows:

$$\tilde{c}_{ij}(\mathbf{x}) t_j(\mathbf{x}) = \int_S \left[K_{ij}(\mathbf{x}, \mathbf{y}) t_j(\mathbf{y}) - H_{ij}(\mathbf{x}, \mathbf{y}) u_j(\mathbf{y}) \right] dS(\mathbf{y})$$
$$+ \int_V K_{ij}(\mathbf{x}, \mathbf{y}) f_j(\mathbf{y}) dV(\mathbf{y}), \quad \forall \mathbf{x} \in S, \tag{4.23}$$

where the coefficients $\tilde{c}_{ij} = 1/2 \delta_{ij}$ if S is smooth at source point **x**. For 2D (plane-strain) problems, the two new kernels are:

$$K_{ij}(\mathbf{x}, \mathbf{y}) = \frac{1}{4\pi(1-v)r} \left[(1-2v)(\delta_{ij} r_{,k} + \delta_{jk} r_{,i} - \delta_{ik} r_{,j}) + 2 r_{,i} r_{,j} r_{,k} \right] n_k(\mathbf{x}),$$
$$(4.24)$$

$$H_{ij}(\mathbf{x}, \mathbf{y}) = \frac{\mu}{2\pi(1-v)r^2} \left\{ 2\frac{\partial r}{\partial n} \left[(1-2v)\delta_{ik} r_{,j} + v(\delta_{ij} r_{,k} + \delta_{jk} r_{,i}) - 4 r_{,i} r_{,j} r_{,k} \right] \right.$$
$$+ 2v(n_i r_{,j} r_{,k} + n_k r_{,i} r_{,j}) - (1-4v)\delta_{ik} n_j$$
$$\left. + (1-2v)(2n_j r_{,i} r_{,k} + \delta_{ij} n_k + \delta_{jk} n_i) \right\} n_k(\mathbf{x}), \tag{4.25}$$

where $n_i(\mathbf{x})$ is the normal at the source point **x**. For 3D problems, the two new kernels are:

$$K_{ij}(\mathbf{x}, \mathbf{y}) = \frac{1}{8\pi(1-v)r^2} \left[(1-2v)(\delta_{ij} r_{,k} + \delta_{jk} r_{,i} - \delta_{ik} r_{,j}) + 3 r_{,i} r_{,j} r_{,k} \right] n_k(\mathbf{x}),$$
$$(4.26)$$

$$H_{ij}(\mathbf{x}, \mathbf{y}) = \frac{\mu}{4\pi(1-v)r^3} \left\{ 3\frac{\partial r}{\partial n} \left[(1-2v)\delta_{ik} r_{,j} + v(\delta_{ij} r_{,k} + \delta_{jk} r_{,i}) - 5 r_{,i} r_{,j} r_{,k} \right] \right.$$
$$+ 3v(n_i r_{,j} r_{,k} + n_k r_{,i} r_{,j}) - (1-4v)\delta_{ik} n_j$$
$$\left. + (1-2v)(3n_j r_{,i} r_{,k} + \delta_{ij} n_k + \delta_{jk} n_i) \right\} n_k(\mathbf{x}). \tag{4.27}$$

In HBIE (4.23), the integral with kernel K is a CPV integral, whereas the one with kernel H is a HFP integral [83, 84]. As in the potential problem case, a dual BIE (or CHBIE) formulation using a linear combination of the CBIE and HBIE can be written as:

$$\text{CBIE} + \beta\text{HBIE} = 0, \tag{4.28}$$

where β is the coupling constant. Dual BIE formulations were found to be very effective and efficient for solving crack problems and problems involving thin shapes [49, 85]. Dual BIE formulations are especially beneficial to the fast multipole BEM because they provide better conditioning for BEM equations and thus can facilitate faster convergence with iterative solvers.

4.4 Weakly Singular Forms of the Boundary Integral Equations

As for the BIEs for potential problems, CBIE (4.19) and HBIE (4.23) can be recast into forms that involve only weakly singular integrals [46–48] or even nonsingular forms without any singular integrals [47]. For example, by using the result in (4.20) for the coefficient $c_{ij}(\mathbf{x})$ in CBIE (4.19), we obtain the following weakly singular form of the CBIE for elastostatics:

$$\gamma u_i(\mathbf{x}) + \int_S T_{ij}(\mathbf{x}, \mathbf{y}) \left[u_j(\mathbf{y}) - u_j(\mathbf{x}) \right] dS(\mathbf{y})$$
$$= \int_S U_{ij}(\mathbf{x}, \mathbf{y}) t_j(\mathbf{y}) dS(\mathbf{y}) + \int_V U_{ij}(\mathbf{x}, \mathbf{y}) f_j(\mathbf{y}) dV(\mathbf{y}), \quad \forall \mathbf{x} \in S, \tag{4.29}$$

in which $\gamma = 0$ for finite domain problems and $\gamma = 1$ for infinite domain problems. The integral with the T kernel is now weakly singular, because:

$$T_{ij}(\mathbf{x}, \mathbf{y}) \left[u_j(\mathbf{y}) - u_j(\mathbf{x}) \right] \sim \begin{cases} O\left(\dfrac{1}{r}\right) O(r) = O(1), & \text{for two dimensions} \\[2mm] O\left(\dfrac{1}{r^2}\right) O(r) = O\left(\dfrac{1}{r}\right), & \text{for three dimensions} \end{cases}$$

as $r \to 0$ if the displacement u_i is continuous.

Similarly, by using the first three identities (4.13)–(4.15) for the fundamental solution, we can derive the following weakly singular form of the HBIE for elastostatics [52]:

$$\gamma t_i(\mathbf{x}) + \int_S H_{ij}(\mathbf{x}, \mathbf{y}) \left[u_j(\mathbf{y}) - u_j(\mathbf{x}) - \frac{\partial u_j}{\partial \xi_\alpha}(\mathbf{x})(\xi_\alpha - \xi_{o\alpha}) \right] dS(\mathbf{y})$$
$$+ E_{jkpq} e_{\alpha q} \frac{\partial u_p}{\partial \xi_\alpha}(\mathbf{x}) \int_S \left[K_{ij}(\mathbf{x}, \mathbf{y}) n_k(\mathbf{y}) + T_{ji}(\mathbf{x}, \mathbf{y}) n_k(\mathbf{x}) \right] dS(\mathbf{y})$$
$$= \int_S \left[K_{ij}(\mathbf{x}, \mathbf{y}) + T_{ji}(\mathbf{x}, \mathbf{y}) \right] t_j(\mathbf{y}) dS(\mathbf{y}) \tag{4.30}$$
$$- \int_S T_{ji}(\mathbf{x}, \mathbf{y}) \left[t_j(\mathbf{y}) - t_j(\mathbf{x}) \right] dS(\mathbf{y}) + \int_\mathbf{V} \mathbf{K}_{ij}(\mathbf{x}, \mathbf{y}) \mathbf{f_j}(\mathbf{y}) d\mathbf{V}(\mathbf{y}), \quad \forall \mathbf{x} \in S,$$

in which ξ_α and $\xi_{o\alpha}$ are the coordinates of $\mathbf{y}$ and $\mathbf{x}$, respectively, in tangential directions ($\alpha = 1$ for two dimensions and $\alpha = 1, 2$ for three dimensions) in the local (natural) coordinate system on an element and $e_{\alpha k} = \partial \xi_\alpha / \partial x_k$ [52]. All the integrals in (4.30) are now, at most, weakly singular if the displacement field u_i has continuous first derivatives.

Weakly singular forms of the BIEs, or regularized BIEs, which do not contain any strongly singular and hypersingular integrals, are useful in cases in which higher-order boundary elements are applied to solve the BIEs. In these cases, analytical evaluations of the singular integrals are difficult or impossible to obtain, and the use of numerical integration is troublesome. When constant elements are used, all the singular and hypersingular integrals can be evaluated analytically (see Appendix A.2 for 2D cases); therefore, the original singular forms of CBIE (4.19) and HBIE (4.23) can be applied directly.

4.5 Discretization of the Boundary Integral Equations

Discretization of the BIEs for elasticity problems is similar to that for the potential problems. The only difference is that we have two or three unknowns at each node for 2D or 3D problems, respectively. For example, the discretized form of CBIE (4.19) can be written as follows (without considering the body force):

$$
\begin{bmatrix}
\mathbf{T}_{11} & \mathbf{T}_{12} & \cdots & \mathbf{T}_{1N} \\
\mathbf{T}_{21} & \mathbf{T}_{22} & \cdots & \mathbf{T}_{2N} \\
\vdots & \vdots & \ddots & \vdots \\
\mathbf{T}_{N1} & \mathbf{T}_{N2} & \cdots & \mathbf{T}_{NN}
\end{bmatrix}
\begin{Bmatrix}
\mathbf{u}_1 \\
\mathbf{u}_2 \\
\vdots \\
\mathbf{u}_N
\end{Bmatrix}
=
\begin{bmatrix}
\mathbf{U}_{11} & \mathbf{U}_{12} & \cdots & \mathbf{U}_{1N} \\
\mathbf{U}_{21} & \mathbf{U}_{22} & \cdots & \mathbf{U}_{2N} \\
\vdots & \vdots & \ddots & \vdots \\
\mathbf{U}_{N1} & \mathbf{U}_{N2} & \cdots & \mathbf{U}_{NN}
\end{bmatrix}
\begin{Bmatrix}
\mathbf{t}_1 \\
\mathbf{t}_2 \\
\vdots \\
\mathbf{t}_N
\end{Bmatrix},
$$
(4.31)

in which $\mathbf{u}_i$ and $\mathbf{t}_i$ are the displacement and traction vectors at node i on boundary S ($i = 1, 2, \ldots, N$), and $\mathbf{T}_{ij}$ and $\mathbf{U}_{ij}$ are 2×2 (for 2D) or 3×3 (for 3D) submatrices we obtain by integrating the T and U kernels, respectively, when the source point $\mathbf{x}$ is at node i and integrations are done on all elements surrounding node j. For 2D constant elements, all the integrals can be evaluated analytically (Appendix A.2), whereas for linear and quadratic elements, numerical integrations need to be used. The diagonal submatrices $\mathbf{T}_{ii}$ can be determined by imposing a rigid-body motion on Eq. (4.31) to obtain:

$$
\mathbf{T}_{ii} =
\begin{cases}
\displaystyle -\sum_{\substack{j \neq i}}^{N} \mathbf{T}_{ij}, & \text{for a finite domain} \\[2mm]
\displaystyle \mathbf{I} - \sum_{\substack{j \neq i}}^{N} \mathbf{T}_{ij}, & \text{for an infinite domain.}
\end{cases}
$$
(4.32)

We can also prove this result by discretizing the weakly singular form of the CBIE in Eq. (4.29) directly [46].

A standard linear system of equations is formed as follows by applying the BC at each node and switching the columns in the two matrices in Eq. (4.31):

$$
\begin{bmatrix}
\mathbf{A}_{11} & \mathbf{A}_{12} & \cdots & \mathbf{A}_{1N} \\
\mathbf{A}_{21} & \mathbf{A}_{22} & \cdots & \mathbf{A}_{2N} \\
\vdots & \vdots & \ddots & \vdots \\
\mathbf{A}_{N1} & \mathbf{A}_{N2} & \cdots & \mathbf{A}_{NN}
\end{bmatrix}
\begin{Bmatrix}
\lambda_1 \\ \lambda_2 \\ \vdots \\ \lambda_N
\end{Bmatrix}
=
\begin{Bmatrix}
\mathbf{b}_1 \\ \mathbf{b}_2 \\ \vdots \\ \mathbf{b}_N
\end{Bmatrix},
\quad \text{or} \quad \mathbf{A}\boldsymbol{\lambda} = \mathbf{b}, \qquad (4.33)
$$

where $\mathbf{A}$ is the coefficient matrix of dimensions $2N \times 2N$ (for two dimensions) or $3N \times 3N$ (for three dimensions), $\boldsymbol{\lambda}$ is the unknown vector, and $\mathbf{b}$ is the known right-hand-side vector (which may also contain contributions from the body forces). Again, the construction of matrix $\mathbf{A}$ requires $O(N^2)$ operations, and the size of the required memory for storing $\mathbf{A}$ is also $O(N^2)$ because $\mathbf{A}$ is, in general, a nonsymmetric and dense matrix. The solution of the system in Eq. (4.33) by use of direct solvers such as Gauss elimination requires $O(N^3)$ operations. Thus, the conventional BEM approach by solving Eq. (4.33) directly is limited to BEM models with only a few thousand equations on a desktop computer. In later sections, we discuss how to apply iterative solvers to the linear system of equations in (4.33) and how to use the FMM to evaluate the far-field contributions in the matrix–vector multiplication in order to accelerate the solutions of the BEM equations for elasticity problems and to achieve the $O(N)$ efficiency.

4.6 Recovery of the Full Stress Field on the Boundary

In stress analysis, values of all the stress components on the boundary are of interest. However, in the BEM solution, only the displacement and traction components on the boundary are solved. The full stress field is not known from this solution, and the most important stress component – for example, the hoop stress on the edge of a hole – is often missing. In the following, we discuss how to recover the full stress field from the BEM solution of the displacement and traction fields, using the 2D case as an example.

For 2D elasticity, we know the values of the displacement components u and v and the traction components t_x and t_y at each node on the boundary after we solve the BEM system of equations. To recover the stress components σ_x, σ_y, and τ_{xy} on the boundary, we proceed as follows.

First, we note the following two equations relating the stress and traction components:

$$
\sigma_x n_x + \tau_{xy} n_y = t_x, \qquad (4.34)
$$

$$
\tau_{xy} n_x + \sigma_y n_y = t_y, \qquad (4.35)
$$

in which n_x and n_y are the direction cosines of the normal n.

Second, we take the derivatives of the displacement field in the tangential direction ξ (local coordinate) of the boundary S to obtain two more relations:

$$\frac{\partial u}{\partial x}\frac{\partial x}{\partial \xi} + \frac{\partial u}{\partial y}\frac{\partial y}{\partial \xi} = \frac{\partial u}{\partial \xi}, \tag{4.36}$$

$$\frac{\partial v}{\partial x}\frac{\partial x}{\partial \xi} + \frac{\partial v}{\partial y}\frac{\partial y}{\partial \xi} = \frac{\partial v}{\partial \xi}, \tag{4.37}$$

where we can readily compute the values of $\partial u/\partial \xi$, $\partial v/\partial \xi$, $\partial x/\partial \xi$, and $\partial y/\partial \xi$ on a boundary element by using the shape functions. For constant elements, we can first compute the averaged displacement values at the end points of all the elements and then apply a linear interpolation to compute the values of these derivatives at the nodes (centers) of the elements.

Third, we write the 2D stress–strain relations as follows:

$$\sigma_x - C\left((1-v)\frac{\partial u}{\partial x} + v\frac{\partial v}{\partial y}\right) = 0, \tag{4.38}$$

$$\sigma_y - C\left(v\frac{\partial u}{\partial x} + (1-v)\frac{\partial v}{\partial y}\right) = 0, \tag{4.39}$$

$$\tau_{xy} - G\left(\frac{\partial u}{\partial y} + \frac{\partial v}{\partial x}\right) = 0, \tag{4.40}$$

where $C = E/[(1+v)(1-2v)]$, $G = E/[2(1+v)]$, E is Young's modulus, and v is Poisson's ratio for the plane-strain case.

Therefore, we have seven equations, Eqs. (4.34)–(4.40), for seven unknowns on the boundary, $\sigma_x, \sigma_y, \tau_{xy}, \partial u/\partial x, \partial u/\partial y, \partial v/\partial x$, and $\partial v/\partial y$, which are sufficient to recover all the stress components on the boundary. Note that the four derivatives of the displacement components $\partial u/\partial x, \partial u/\partial y, \partial v/\partial x$, and $\partial v/\partial y$ can be used to determine directly the strain components $\varepsilon_x, \varepsilon_y$, and γ_{xy}, if needed.

Combining the seven equations in (4.34)–(4.40), we obtain the following linear system of equations for the recovery of the full stresses (and strains) in the 2D case:

$$
\begin{bmatrix}
n_x & 0 & n_y & 0 & 0 & 0 & 0 \\
0 & n_y & n_x & 0 & 0 & 0 & 0 \\
0 & 0 & 0 & \frac{\partial x}{\partial \xi} & \frac{\partial y}{\partial \xi} & 0 & 0 \\
0 & 0 & 0 & 0 & 0 & \frac{\partial x}{\partial \xi} & \frac{\partial y}{\partial \xi} \\
1 & 0 & 0 & -C(1-v) & 0 & 0 & -Cv \\
0 & 1 & 0 & -Cv & 0 & 0 & -C(1-v) \\
0 & 0 & 1 & 0 & -G & -G & 0
\end{bmatrix}
\left\{
\begin{array}{c}
\sigma_x \\
\sigma_y \\
\tau_{xy} \\
\frac{\partial u}{\partial x} \\
\frac{\partial u}{\partial y} \\
\frac{\partial v}{\partial x} \\
\frac{\partial v}{\partial y}
\end{array}
\right\}
=
\left\{
\begin{array}{c}
t_x \\
t_y \\
\frac{\partial u}{\partial \xi} \\
\frac{\partial v}{\partial \xi} \\
0 \\
0 \\
0
\end{array}
\right\}. \tag{4.41}
$$

We can derive a similar linear system with 15 equations for the 3D elasticity case by following the same approach, which can be applied to recover all the six stress (and strain) components on the boundary surface.

4.7 Fast Multipole Boundary Element Method for 2D Elastostatic Problems

The fast multipole algorithms for solving general 2D elasticity problems by using CBIE (4.19) and HBIE (4.23) are described in detail in this section. As in the 2D potential case, complex notation is used. The kernels are represented by complex functions from the classical 2D elasticity theory.

First, we note that the two integrals in CBIE (4.19) can be represented in complex variables readily if we write the fundamental solution $U_{ij}(\mathbf{x}, \mathbf{y})$ and $T_{ij}(\mathbf{x}, \mathbf{y})$ in the complex notation by using the results in 2D elasticity. In 2D elasticity theory with complex variables, the displacement field $U = U_1 + iU_2$ at a field point $z(= y_1 + iy_2$, with $i = \sqrt{-1})$ because of a point force $P = P_1 + iP_2$ at the source point $z_0(= x_1 + ix_2)$ can be written as (see, e.g., Refs. [86, 87]):

$$U_1(z) + iU_2(z) = \frac{1}{4\pi\mu(1+\kappa)}\left\{-\kappa P\left[\log(z_0 - z) + \overline{\log(z_0 - z)}\right] + \overline{P}\frac{z_0 - z}{\overline{z_0 - z}}\right\},$$
(4.42)

in which the overbar indicates the complex conjugate and $\kappa = 3 - 4\nu$ for the plane-strain case.

We can obtain the fundamental solution U_{ij} exactly as given in Eq. (4.9) by letting $P = 1$ and i (first in the x direction, then in the y direction, respectively), in Eq. (4.42). Using the preceding result, we can show that the first integral in CBIE (4.19) can be written in the following complex form by applying Eq. (4.42) (with no body force) [73]:

$$\frac{1}{2}u(z_0) = D_t(z_0) - D_u(z_0),$$
(4.43)

where $u = u_1 + iu_2$ is the complex representation of the displacement field and boundary S is assumed to be smooth at the source point z_0. In the preceding equation:

$$D_t(z_0) \equiv \left[\int_S U_{1j}(\mathbf{x}, \mathbf{y})t_j(\mathbf{y})dS(\mathbf{y})\right] + i\left[\int_S U_{2j}(\mathbf{x}, \mathbf{y})t_j(\mathbf{y})dS(\mathbf{y})\right]$$
$$= \frac{1}{2\mu(1+\kappa)}\int_S\left[\kappa G(z_0, z)t(z) - (z_0 - z)\overline{G'(z_0, z)}\,\overline{t(z)}\right.$$
$$\left. + \kappa\overline{G(z_0, z)}t(z)\right]dS(z),$$
(4.44)

representing the first integral with the U kernel in CBIE (4.19), and:

$$
\begin{aligned}
D_u(z_0) &\equiv \left[\int_S T_{1j}(\mathbf{x}, \mathbf{y}) u_j(\mathbf{y}) dS(\mathbf{y}) \right] + i \left[\int_S T_{2j}(\mathbf{x}, \mathbf{y}) u_j(\mathbf{y}) dS(\mathbf{y}) \right] \\
&= -\frac{1}{1+\kappa} \int_S \left\{ \kappa\, G'(z_0, z) n(z) u(z) - (z_0 - z) \overline{G''(z_0, z)}\, \overline{n(z)}\, \overline{u(z)} \right. \\
&\qquad \left. + \overline{G'(z_0, z)} \left[n(z) \overline{u(z)} + \overline{n(z)} u(z) \right] \right\} dS(z),
\end{aligned} \tag{4.45}
$$

representing the second integral with the T kernel in CBIE (4.19), where $t = t_1 + it_2$ and $n = n_1 + in_2$ are the complex traction and normal, respectively:

$$
G(z_0, z) = -\frac{1}{2\pi} \log(z_0 - z) \tag{4.46}
$$

is the Green's function (in complex form) for 2D potential problems [see Eq. (3.7)], and $(\ )' \equiv \partial(\ )/\partial z_0$.

To derive the complex form of HBIE (4.23), we first note that the real variable traction t_i on boundary S is given by:

$$
t_i = \sigma_{ij} n_j = [\lambda \delta_{ij} u_{k,k} + \mu(u_{i,j} + u_{j,i})] n_j, \tag{4.47}
$$

in which σ_{ij} is the stress tensor and $\lambda = 2\mu\nu/(1 - 2\nu)$ for plane-strain problems. It is interesting to note that this relation can be written in complex form as follows:

$$
t(z) = 2\mu \left[\frac{1}{\kappa - 1} \left(\frac{\partial u}{\partial z} + \frac{\partial \overline{u}}{\partial \overline{z}} \right) n + \frac{\partial u}{\partial \overline{z}} \overline{n} \right], \tag{4.48}
$$

in which $t, u,$ and n are the complex traction, displacement, and normal on boundary S, respectively. In applying this formula, z and $\overline{z}$ must be considered as two independent variables; that is, $\partial z/\partial \overline{z} = \partial \overline{z}/\partial z = 0$. It is straightforward to verify that Eq. (4.48) is indeed equivalent to Eq. (4.47) by simply extracting the real and imaginary parts of $t(z)$ from Eq. (4.48) and comparing with the results we obtain by expanding Eq. (4.47).

Applying the relation in Eq. (4.48), we can show that HBIE (4.23) can be written in the following complex form (with no body force):

$$
\frac{1}{2} t(z_0) = F_t(z_0) - F_u(z_0), \tag{4.49}
$$

where:

$$
F_t(z_0) = 2\mu \left[\frac{1}{\kappa - 1} \left(\frac{\partial D_t(z_0)}{\partial z_0} + \frac{\partial \overline{D_t(z_0)}}{\partial \overline{z_0}} \right) n(z_0) + \frac{\partial D_t(z_0)}{\partial \overline{z_0}} \overline{n(z_0)} \right] \tag{4.50}
$$

represents the first integral with the K kernel in HBIE (4.23), and:

$$
F_u(z_0) = 2\mu \left[\frac{1}{\kappa - 1} \left(\frac{\partial D_u(z_0)}{\partial z_0} + \frac{\partial \overline{D_u(z_0)}}{\partial \overline{z_0}} \right) n(z_0) + \frac{\partial D_u(z_0)}{\partial \overline{z_0}} \overline{n(z_0)} \right] \tag{4.51}
$$

represents the second integral with the H kernel in HBIE (4.23). Applying Eqs. (4.44) and (4.45), we obtain the following explicit results:

$$
\begin{aligned}
F_t(z_0) &\equiv [F_1(\mathbf{x}) + i F_2(\mathbf{x})]_t \\
&\equiv \left[\int_S K_{1j}(\mathbf{x}, \mathbf{y}) t_j(\mathbf{y}) dS(\mathbf{y}) \right] + i \left[\int_S K_{2j}(\mathbf{x}, \mathbf{y}) t_j(\mathbf{y}) dS(\mathbf{y}) \right] \\
&= \frac{1}{1 + \kappa} \int_S \{ \, [G'(z_0, z) t(z) + \overline{G'(z_0, z) t(z)}] \, n(z_0) \\
&\quad + [\kappa \overline{G'(z_0, z)} t(z) - (z_0 - z) \overline{G''(z_0, z) t(z)}] \, \overline{n(z_0)} \, \} dS(z),
\end{aligned} \tag{4.52}
$$

$$
\begin{aligned}
F_u(z_0) &\equiv [F_1(\mathbf{x}) + i F_2(\mathbf{x})]_u \\
&\equiv \left[\int_S H_{1j}(\mathbf{x}, \mathbf{y}) u_j(\mathbf{y}) dS(\mathbf{y}) \right] + i \left[\int_S H_{2j}(\mathbf{x}, \mathbf{y}) u_j(\mathbf{y}) dS(\mathbf{y}) \right] \\
&= -\frac{2\mu}{1 + \kappa} \int_S (\, [G''(z_0, z) n(z) u(z) + \overline{G''(z_0, z) n(z) u(z)}] \, n(z_0) \\
&\quad + \{ \overline{G''(z_0, z)} [n(z) \overline{u(z)} + \overline{n(z)} u(z)] \\
&\quad - (z_0 - z) \overline{G'''(z_0, z) n(z) u(z)} \} \overline{n(z_0)}) dS(z).
\end{aligned} \tag{4.53}
$$

To show that complex variable CBIE (4.43) is equivalent to real variable CBIE (4.19) and complex variable HBIE (4.49) is equivalent to real variable HBIE (4.23), we can simply introduce the polar coordinate system (r, θ) with the origin at z_0; notice that:

$$
z - z_0 = r e^{i\theta}, \quad G' = \frac{1}{2\pi (z - z_0)} = \frac{1}{2\pi r} e^{-i\theta}, \quad \overline{G'} = \frac{1}{2\pi r} e^{i\theta}, \quad \text{and so on,} \tag{4.54}
$$

and extract the real and imaginary parts of the results in the complex variable BIEs.

In the following discussion, we first study the multipole expansions, local expansions, and their translations related to Eqs. (4.44) and (4.45) in the fast multipole BEM for CBIE (4.43). Then, we present the expansions related to Eqs. (4.52) and (4.53) for HBIE (4.49). The derivations of these results are similar and closely related to those for 2D potential problems discussed in the previous chapter.

4.7.1 Multipole Expansion for the U Kernel Integral

Let z_c be a multipole expansion point close to z (Figure 3.2) – that is, $|z - z_c| \ll |z_0 - z_c|$; the *multipole expansion* for $D_t(z_0)$ in (4.44) with the U

kernel is given by [73]:

$$D_t(z_0) = \frac{1}{4\pi\mu(1+\kappa)}\left[\kappa\sum_{k=0}^{\infty} O_k(z_0-z_c)M_k(z_c) + z_0\sum_{k=0}^{\infty} \overline{O_{k+1}(z_0-z_c)}\,\overline{M_k(z_c)}\right.$$

$$\left. + \sum_{k=0}^{\infty} \overline{O_k(z_0-z_c)N_k(z_c)}\right],$$

(4.55)

where:

$$M_k(z_c) = \int_{S_c} I_k(z-z_c)t(z)dS(z), \quad \text{for } k \geq 0,$$

(4.56)

$$N_0 = \kappa\int_{S_c} t(z)dS(z);$$

$$N_k(z_c) = \int_{S_c}\left[\kappa\overline{I_k(z-z_c)}t(z) - \overline{I_{k-1}(z-z_c)}z\overline{t(z)}\right]dS(z), \quad \text{for } k \geq 1,$$

(4.57)

are the two sets of *moments* about z_c, with S_c being a subset of S that is far away from the source point z_0 (Figure 3.2). The two auxiliary functions $I_k(z)$ and $O_k(z)$ were defined in Eqs. (3.12). Equation (4.55) is derived readily by use of the expansion for $G(z_0, z)$ given in Eq. (3.11).

4.7.2 Moment-to-Moment Translation

If the multipole expansion point z_c is moved to a new location $z_{c'}$ (Figure 3.2), we have:

$$M_k(z_{c'}) = \sum_{l=0}^{k} I_{k-l}(z_c - z_{c'})M_l(z_c), \quad \text{for } k \geq 0.$$

(4.58)

Similarly,

$$N_k(z_{c'}) = \sum_{l=0}^{k} \overline{I_{k-l}(z_c - z_{c'})}N_l(z_c), \quad \text{for } k \geq 0.$$

(4.59)

These are the *M2M translations* for the moments when z_c is moved to $z_{c'}$. Note that these translation coefficients are symmetrical for the two sets of moments (I_{k-l} and conjugate of I_{k-l}) and coefficients I_{k-l} are exactly the same as used in the 2-D potential case (see Eq. (3.23)).

4.7.3 Local Expansion and Moment-to-Local Translation

Let z_L be a local expansion point close to the source point z_0 (Figure 3.2); that is, $|z_0 - z_L| \ll |z_c - z_L|$. Expanding $D_t(z_0)$ in (4.55) about $z_0 = z_L$ by using a

Taylor series expansion, we have the following *local expansion* [73]:

$$D_t(z_0) = \frac{1}{4\pi\mu(1+\kappa)} \left[\kappa \sum_{l=0}^{\infty} L_l(z_L) I_l(z_0 - z_L) - z_0 \sum_{l=1}^{\infty} \overline{L_l(z_L) I_{l-1}(z_0 - z_L)} \right.$$

$$\left. + \sum_{l=0}^{\infty} K_l(z_L) \overline{I_l(z_0 - z_L)} \right], \tag{4.60}$$

where the coefficients are given by the following M2L translations:

$$L_l(z_L) = (-1)^l \sum_{k=0}^{\infty} O_{l+k}(z_L - z_c) M_k(z_c), \quad \text{for } l \geq 0; \tag{4.61}$$

$$K_l(z_L) = (-1)^l \sum_{k=0}^{\infty} \overline{O_{l+k}(z_L - z_c)} N_k(z_c), \quad \text{for } l \geq 0. \tag{4.62}$$

Note that these M2L translation coefficients are also symmetrical regarding the translation coefficients [see Eq. (3.25)].

4.7.4 Local-to-Local Translation

If the local expansion point is moved from z_L to $z_{L'}$ (Figure 3.2), the new local expansion coefficients are given by the following *L2L translations* [73]:

$$L_l(z_{L'}) = \sum_{m=l}^{\infty} I_{m-l}(z_{L'} - z_L) L_m(z_L), \quad \text{for } l \geq 0; \tag{4.63}$$

$$K_l(z_{L'}) = \sum_{m=l}^{\infty} \overline{I_{m-l}(z_{L'} - z_L)} K_m(z_L), \quad \text{for } l \geq 0, \tag{4.64}$$

which are also symmetrical regarding the translation coefficients [see Eq. (3.30)].

4.7.5 Expansions for the *T* Kernel Integral

Through a procedure similar to that used for the U kernel integral in (4.44), the multipole expansion of the T kernel integral $D_u(z_0)$ in (4.45) can be written as [73]:

$$D_u(z_0) = \frac{1}{2\pi(1+\kappa)} \left[\kappa \sum_{k=1}^{\infty} O_k(z_0 - z_c) \tilde{M}_k(z_c) + z_0 \sum_{k=1}^{\infty} \overline{O_{k+1}(z_0 - z_c) \tilde{M}_k(z_c)} \right.$$

$$\left. + \sum_{k=1}^{\infty} \overline{O_k(z_0 - z_c)} \tilde{N}_k(z_c) \right], \tag{4.65}$$

where the two sets of *moments* are:

$$\tilde{M}_k(z_c) = \int_{S_c} I_{k-1}(z - z_c)n(z)u(z)dS(z), \quad \text{for } k \geq 1; \qquad (4.66)$$

$$\tilde{N}_1 = \int_{S_c} \left[n(z)\overline{u(z)} + \overline{n(z)}u(z) \right] dS(z);$$

$$\tilde{N}_k(z_c) = \int_{S_c} \left\{ \overline{I_{k-1}(z - z_c)} \left[n(z)\overline{u(z)} + \overline{n(z)}u(z) \right] \right. \qquad (4.67)$$
$$\left. - \overline{I_{k-2}(z - z_c)} \, \overline{zn(z)u(z)} \right\} dS(z), \quad \text{for } k \geq 2.$$

These moments are similar to those for the U kernel integral. It can be shown that all the M2M, M2L, and L2L translations remain the same for the T kernel integrals, except that $\tilde{M}_0 = \tilde{N}_0 = 0$. In fact, moments M_k and $\tilde{M}_k$ are combined, as well as moments N_k and $\tilde{N}_k$, so that only two sets of moments are involved in the M2M and M2L translations.

The local expansion for $D_u(z_0)$ is [73]:

$$D_u(z_0) = \frac{1}{2\pi(1 + \kappa)} \left[\kappa \sum_{l=0}^{\infty} L_l(z_L)I_l(z_0 - z_L) - z_0 \sum_{l=1}^{\infty} \overline{L_l(z_L)I_{l-1}(z_0 - z_L)} \right.$$
$$\left. + \sum_{l=0}^{\infty} K_l(z_L)\overline{I_l(z_0 - z_L)} \right], \qquad (4.68)$$

where the local expansion coefficients $L_l(z_L)$ and $K_l(z_L)$ are given by Eqs. (4.61) and (4.62), with M_k and N_k replaced with $\tilde{M}_k$ and $\tilde{N}_k$, respectively.

4.7.6 Expansions for the Hypersingular Boundary Integral Equation

To derive the multipole expansions and local expansions for HBIE (4.49), we can simply take the derivatives of the local expansions for the two integrals in the CBIE – that is, the integrals in Eqs. (4.60) and (4.68), respectively – and then invoke the constitutive relation in the complex form; that is, Eqs. (4.50) and (4.51). The result of the local expansion for the first integral $F_t(z_0)$ in Eq. (4.52) for the HBIE is:

$$F_t(z_0) = \frac{1}{2\pi(1 + \kappa)} \left\{ \left[\sum_{l=0}^{\infty} L_{l+1}(z_L)I_l(z_0 - z_L) + \sum_{l=0}^{\infty} \overline{L_{l+1}(z_L)I_l(z_0 - z_L)} \right] n(z_0) \right.$$
$$\left. + \left[-z_0 \sum_{l=1}^{\infty} \overline{L_{l+1}(z_L)I_{l-1}(z_0 - z_L)} + \sum_{l=0}^{\infty} K_{l+1}(z_L)\overline{I_l(z_0 - z_L)} \right] \overline{n(z_0)} \right\},$$
$$(4.69)$$

in which the expansion coefficients $L_l(z_L)$ and $K_l(z_L)$ are given by the same M2L translations in (4.61) and (4.62), respectively. That is, the same sets of moments M_k and N_k used for $D_t(z_0)$ are used for $F_t(z_0)$ directly.

Similarly, it can be shown that the local expansion for the second integral $F_u(z_0)$ in Eq. (4.53) for the HBIE is:

$$F_u(z_0) = \frac{\mu}{\pi(1+\kappa)} \left\{ \left[\sum_{l=0}^{\infty} L_{l+1}(z_L) I_l(z_0 - z_L) + \sum_{l=0}^{\infty} \overline{L_{l+1}(z_L) I_l(z_0 - z_L)} \right] n(z_0) \right.$$
$$\left. + \left[-z_0 \sum_{l=1}^{\infty} \overline{L_{l+1}(z_L) I_{l-1}(z_0 - z_L)} + \sum_{l=0}^{\infty} K_{l+1}(z_L) \overline{I_l(z_0 - z_L)} \right] \overline{n(z_0)} \right\},$$

$$(4.70)$$

in which $L_l(z_L)$ and $K_l(z_L)$ are given by Eqs. (4.61) and (4.62), with M_k replaced by $\tilde{M}_k$ and N_k with $\tilde{N}_k$. Again, the same sets of moments $\tilde{M}_k$ and $\tilde{N}_k$ used for $D_u(z_0)$ are used for $F_u(z_0)$ directly; thus, all of the M2M, M2L, and L2L translations for the HBIE remain the same as those used for the CBIE.

The details of the fast multipole algorithms for solving 2D elasticity problems are similar to those for 2D potential problems, which are described in the previous chapter. For example, if constant boundary elements (straight-line segment with one node) are applied to discretize the 2D elasticity BIEs, all of the moments can be evaluated analytically, as well as the integration of the kernels in the near-field direct evaluations (Appendix A.2).

4.8 Fast Multipole Boundary Element Method for 3D Elastostatic Problems

To discuss the fast multipole formulation for 3D elasticity problems, we first note that the fundamental solution in Eq. (4.11) can be written in the following form:

$$U_{ij}(\mathbf{x}, \mathbf{y}) = \frac{1}{8\pi\mu} \left(\delta_{ij} \frac{2}{r} - \frac{\lambda+\mu}{\lambda+2\mu} \frac{\partial}{\partial x_i} \frac{x_j - y_j}{r} \right). \quad (4.71)$$

Start with the following expansion [see Eq. (3.46) used for 3-D potential problems]:

$$\frac{1}{r(\mathbf{x}, \mathbf{y})} = \sum_{n=0}^{\infty} \sum_{m=-n}^{n} \overline{S_{n,m}(\mathbf{x} - \mathbf{y}_c)} R_{n,m}(\mathbf{y} - \mathbf{y}_c), \quad |\mathbf{y} - \mathbf{y}_c| < |\mathbf{x} - \mathbf{y}_c|, \quad (4.72)$$

where $\mathbf{y}_c$ is an expansion point close to the field point $\mathbf{y}$ and the overbar indicates the complex conjugate. The functions $R_{n,m}$ and $S_{n,m}$ are solid harmonic functions given in Eqs. (3.47) and (3.48), respectively. Note that the left-hand side of Eq. (4.72) is a real function. Therefore, the complex conjugate can also be placed on $R_{n,m}$ in Eq. (4.72). Substituting the results in (4.72) into Eq. (4.71), we arrive at:

$$U_{ij}(\mathbf{x}, \mathbf{y}) = \frac{1}{8\pi\mu} \sum_{n=0}^{\infty} \sum_{m=-n}^{n} \left[F_{ij,n,m}(\mathbf{x} - \mathbf{y}_c) \overline{R_{n,m}(\mathbf{y} - \mathbf{y}_c)} \right.$$
$$\left. + G_{i,n,m}(\mathbf{x} - \mathbf{y}_c)(\mathbf{y} - \mathbf{y}_c)_j \overline{R_{n,m}(\mathbf{y} - \mathbf{y}_c)} \right], \quad (4.73)$$

where:

$$F_{ij,n,m}(\mathbf{x} - \mathbf{y}_c) \equiv \frac{\lambda + 3\mu}{\lambda + 2\mu} \delta_{ij} S_{n,m}(\mathbf{x} - \mathbf{y}_c) - \frac{\lambda + \mu}{\lambda + 2\mu} (\mathbf{x} - \mathbf{y}_c)_j \frac{\partial}{\partial x_i} S_{n,m}(\mathbf{x} - \mathbf{y}_c),$$

(4.74)

$$G_{i,n,m}(\mathbf{x} - \mathbf{y}_c) \equiv \frac{\lambda + \mu}{\lambda + 2\mu} \frac{\partial}{\partial x_i} S_{n,m}(\mathbf{x} - \mathbf{y}_c).$$

(4.75)

Consider the first integral with the U kernel in CBIE (4.19) on a subdomain S_c of S away from the source point $\mathbf{x}$. Applying expression (4.73), with point $\mathbf{y}_c$ being close to subdomain S_c (elements within a leaf), we obtain the following *multipole expansion*:

$$\int_{S_c} U_{ij}(\mathbf{x}, \mathbf{y}) t_j(\mathbf{y}) dS(\mathbf{y}) = \frac{1}{8\pi\mu} \sum_{n=0}^{\infty} \sum_{m=-n}^{n} \left[F_{ij,n,m}(\mathbf{x} - \mathbf{y}_c) \overline{M_{j,n,m}(\mathbf{y}_c)} \right.$$

$$\left. + G_{i,n,m}(\mathbf{x} - \mathbf{y}_c) \overline{M_{n,m}(\mathbf{y}_c)} \right],$$

(4.76)

in which:

$$M_{j,n,m}(\mathbf{y}_c) = \int_{S_c} R_{n,m}(\mathbf{y} - \mathbf{y}_c) t_j(\mathbf{y}) dS(\mathbf{y}),$$

$$M_{n,m}(\mathbf{y}_c) = \int_{S_c} (\mathbf{y} - \mathbf{y}_c)_j R_{n,m}(\mathbf{y} - \mathbf{y}_c) t_j(\mathbf{y}) dS(\mathbf{y})$$

(4.77)

are the *moments* for given n and m. Evaluations of these four moments are independent of the location of the source point $\mathbf{x}$ and thus need to be calculated only once on each element.

To obtain the multipole expansion for the T kernel integral in CBIE (4.19), we note that:

$$T_{ij}(\mathbf{x}, \mathbf{y}) = E_{jklp} n_k(\mathbf{y}) \frac{\partial}{\partial y_p} U_{il}(\mathbf{x}, \mathbf{y}).$$

(4.78)

From this relation and expansion in Eq. (4.73), we obtain the *multipole expansion* for the T kernel integral as follows:

$$\int_{S_c} T_{ij}(\mathbf{x}, \mathbf{y}) u_j(\mathbf{y}) dS(\mathbf{y}) = \frac{1}{8\pi\mu} \sum_{n=0}^{\infty} \sum_{m=-n}^{n} \left[F_{ij,n,m}(\mathbf{x} - \mathbf{y}_c) \overline{\widetilde{M}_{j,n,m}(\mathbf{y}_c)} \right.$$

$$\left. + G_{i,n,m}(\mathbf{x} - \mathbf{y}_c) \overline{\widetilde{M}_{n,m}(\mathbf{y}_c)} \right],$$

(4.79)

in which:

$$\widetilde{M}_{j,n,m}(\mathbf{y}_c) = E_{jpkl} \int_{S_c} \frac{\partial}{\partial y_p} [R_{n,m}(\mathbf{y} - \mathbf{y}_c)] n_k(\mathbf{y}) u_l(\mathbf{y}) dS(\mathbf{y}),$$

$$\widetilde{M}_{n,m}(\mathbf{y}_c) = E_{jpkl} \int_{S_c} \frac{\partial}{\partial y_p} [(\mathbf{y} - \mathbf{y}_c)_j R_{n,m}(\mathbf{y} - \mathbf{y}_c)] n_k(\mathbf{y}) u_l(\mathbf{y}) dS(\mathbf{y}).$$

(4.80)

Depending on the boundary conditions, only one in each of the pairs $(M_{j,n,m}, \widetilde{M}_{j,n,m})$ and $(M_{n,m}, \widetilde{M}_{n,m})$ are used in the moment calculations. There is a total of four moments that need to calculated on each boundary element.

When the expansion point is moved from $\mathbf{y}_c$ to $\mathbf{y}_{c'}$, we have the following M2M translations:

$$M_{j,n,m}(\mathbf{y}_{c'}) = \sum_{n'=0}^{n} \sum_{m'=-n'}^{n'} R_{n',m'}(\mathbf{y}_{c'} - \mathbf{y}_c) M_{j,n-n',m-m'}(\mathbf{y}_c),$$

$$M_{n,m}(\mathbf{y}_{c'}) = \sum_{n'=0}^{n} \sum_{m'=-n'}^{n'} R_{n',m'}(\mathbf{y}_{c'} - \mathbf{y}_c) \left[M_{n-n',m-m'}(\mathbf{y}_c) \right. \tag{4.81}$$

$$\left. + (\mathbf{y}_{c'} - \mathbf{y}_c)_j M_{j,n-n',m-m'}(\mathbf{y}_c) \right],$$

which are also valid for $\widetilde{M}_{j,n,m}$ and $\widetilde{M}_{n,m}$.

The local expansion of the U kernel integral on S_c about the point $\mathbf{x} = \mathbf{x}_L$ is given as follows:

$$\int_{S_c} U_{ij}(\mathbf{x}, \mathbf{y}) t_j(\mathbf{y}) dS(\mathbf{y}) = \frac{1}{8\pi \mu} \sum_{n=0}^{\infty} \sum_{m=-n}^{n} \left[F_{ij,n,m}^{R}(\mathbf{x} - \mathbf{x}_L) L_{j,n,m}(\mathbf{x}_L) \right.$$

$$\left. + G_{i,n,m}^{R}(\mathbf{x} - \mathbf{x}_L) L_{n,m}(\mathbf{x}_L) \right], \tag{4.82}$$

where the local expansion coefficients are given by the following M2L translations:

$$L_{j,n,m}(\mathbf{x}_L) = (-1)^n \sum_{n'=0}^{\infty} \sum_{m'=-n'}^{n'} \overline{S_{n+n',m+m'}}(\mathbf{x}_L - \mathbf{y}_c) M_{j,n',m'}(\mathbf{y}_c),$$

$$L_{n,m}(\mathbf{x}_L) = (-1)^n \sum_{n'=0}^{\infty} \sum_{m'=-n'}^{n'} \overline{S_{n+n',m+m'}}(\mathbf{x}_L - \mathbf{y}_c) \tag{4.83}$$

$$\times \left[M_{n',m'}(\mathbf{y}_c) - (\mathbf{x}_L - \mathbf{y}_c)_j M_{j,n',m'}(\mathbf{y}_c) \right],$$

and $F_{ij,n,m}^{R}$ and $G_{i,n,m}^{R}$ are obtained from Eqs. (4.74) and (4.75), respectively, with $S_{n,m}$ replaced with $R_{n,m}$ in each case. The local expansion for the T kernel integral is similar to that of Eq. (4.82), only with $M_{j,n',m'}$ and $M_{n',m'}$ replaced with $\widetilde{M}_{j,n',m'}$ and $\widetilde{M}_{n',m'}$, respectively, in Eq. (4.83) when the local expansion coefficients for the T kernel integral are computed.

When the local expansion point is moved from $\mathbf{x}_L$ to $\mathbf{x}_{L'}$, we have the following L2L translations:

$$L_{j,n,m}(\mathbf{x}_{L'}) = \sum_{n'=n}^{\infty} \sum_{m'=-n'}^{n'} R_{n'-n,m'-m}(\mathbf{x}_{L'} - \mathbf{x}_L) L_{j,n',m'}(\mathbf{x}_L),$$

$$\tag{4.84}$$

$$L_{n,m}(\mathbf{x}_{L'}) = \sum_{n'=n}^{\infty} \sum_{m'=-n'}^{n'} R_{n'-n,m'-m}(\mathbf{x}_{L'} - \mathbf{x}_L) \left[L_{n',m'}(\mathbf{x}_L) - (\mathbf{x}_{L'} - \mathbf{x}_L)_j L_{j,n',m'}(\mathbf{x}_L) \right].$$

We can readily obtain the fast multipole formulation for the HBIE by taking the derivatives of the local expansions for the CBIE and invoking the constitutive equations, as we did in the 2D elasticity case.

As in the 2D cases, the fast multipole formulations for 3D elasticity problems closely resemble those for 3D potential problems. In fact, all of the expansions and translations (M2M, M2L, and L2L) for the 3D elasticity case are similar to those for 3D potential cases, as we discussed in the previous chapter. The only difference is that we have four moments in the multipole expansion for 3D elasticity problems, whereas we have only one moment for 3D potential problems. From this fact, we can readily obtain a fast multipole BEM program for 3D elasticity problems by extending a fast multipole BEM program for 3D potential problems.

It should be pointed out that the M2L translations are more expensive compared with other operations in the FMM, especially for 3D vector problems. A new version of the FMM was introduced by Greengard and Rokhlin in 1997 [65], which uses exponential expansions and replaces the M2L translations with multipole-to-exponential (M2X), exponential-to-exponential (X2X), and exponential-to-local (X2L) expansions. This new version is more difficult to implement. However, it can speed up the solutions by about 20%–40% for many 3D applications [61, 66].

4.9 Fast Multipole Boundary Element Method for Multidomain Elasticity Problems

In this section, we discuss a BEM formulation for multidomain elasticity problems [74] that can be applied to model fiber-reinforced composite materials, functionally graded materials, and other inclusion problems in elasticity. The efficient preconditioner for this BEM formulation can be constructed that can provide efficient solution strategies for the fast multipole BEM for solving such multidomain elasticity problems.

Consider a 2D or 3D elastic domain V_0 with boundary S_0 and embedded with n elastic inclusions V_α with interface S_α, where $\alpha = 1, 2, \ldots, n$ (Figure 4.2). In this discussion, we assume all the inclusions are completely embedded inside the elastic matrix domain; that is, there is no intersection of the interface S_α with the outer boundary S_0. This is a special case of the general multidomain problems. We also assume that no body force is present.

For the matrix domain V_0, we have the following CBIE from Eq. (4.19):

$$\frac{1}{2}u_i(\mathbf{x}) = \int_S [U_{ij}(\mathbf{x}, \mathbf{y})t_j(\mathbf{y}) - T_{ij}(\mathbf{x}, \mathbf{y})u_j(\mathbf{y})]dS(\mathbf{y}), \quad \forall \mathbf{x} \in S, \qquad (4.85)$$

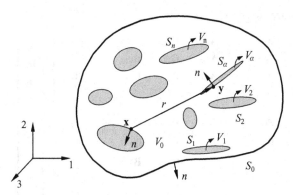

Figure 4.2. Matrix domain V_0 and n inclusions.

where u_i and t_i are the displacement and traction, respectively; $S = \cup_{\alpha=0}^{n} S_\alpha$ is the total boundary of domain V_0 (assuming that S is smooth around $\mathbf{x}$) and $U_{ij}(\mathbf{x}, \mathbf{y})$ and $T_{ij}(\mathbf{x}, \mathbf{y})$ are the two kernel functions.

For each inclusion, the CBIE from Eq. (4.19) can be written as:

$$\frac{1}{2} u_i^{(\alpha)}(\mathbf{x}) = \int_{S_\alpha} \left[U_{ij}^{(\alpha)}(\mathbf{x}, \mathbf{y}) t_j^{(\alpha)}(\mathbf{y}) - T_{ij}^{(\alpha)}(\mathbf{x}, \mathbf{y}) u_j^{(\alpha)}(\mathbf{y}) \right] dS(\mathbf{y}), \quad \forall \mathbf{x} \in S_\alpha, \quad (4.86)$$

for $\alpha = 1, 2, \ldots, n$, in which $u_i^{(\alpha)}$ and $t_i^{(\alpha)}$ are the displacement and traction, respectively, for inclusion α, and $U_{ij}^{(\alpha)}(\mathbf{x}, \mathbf{y})$ and $T_{ij}^{(\alpha)}(\mathbf{x}, \mathbf{y})$ are the two kernels using the shear modulus, Poisson's ratio, and outward normal for inclusion α.

HBIE (4.23) can also be applied in the matrix as well as in the inclusion domains. In fact, the dual BIE formulation (CHBIE, a linear combination of the CBIE and HBIE) is preferred for modeling inclusion problems in which thin shapes often exist and can present difficulties for the CBIE formulation when it is applied alone.

Assume that the inclusions are perfectly bonded to the matrix; that is, there are no gaps or cracks and no interphase regions. We have the following interface conditions:

$$u_i = u_i^{(\alpha)}, \quad t_i = -t_i^{(\alpha)}, \quad (4.87)$$

for $\alpha = 1, 2, \ldots, n$, which state that the displacements are continuous and the tractions are in equilibrium at the interfaces.

From the assumptions just mentioned, we can write the discretized form of the multidomain BIEs by using either the CBIE or CHBIE for the matrix domain and the inclusions as follows [74]:

$$
\begin{aligned}
&\text{Matrix}
\begin{cases}
S_0 \\ S_1 \\ S_2 \\ \vdots \\ S_n
\end{cases} \\
&\text{Inclusions}
\begin{cases}
S_1 \\ S_2 \\ \vdots \\ S_n
\end{cases}
\end{aligned}
\begin{bmatrix}
\mathbf{A}_{00} & \mathbf{A}_{01} & \mathbf{A}_{02} & \cdots & \mathbf{A}_{0n} & -\mathbf{B}_{01} & -\mathbf{B}_{02} & \cdots & -\mathbf{B}_{0n} \\
\mathbf{A}_{10} & \mathbf{A}_{11} & \mathbf{A}_{12} & \cdots & \mathbf{A}_{1n} & -\mathbf{B}_{11} & -\mathbf{B}_{12} & \cdots & -\mathbf{B}_{1n} \\
\mathbf{A}_{20} & \mathbf{A}_{21} & \mathbf{A}_{22} & \cdots & \mathbf{A}_{2n} & -\mathbf{B}_{21} & -\mathbf{B}_{22} & \cdots & -\mathbf{B}_{2n} \\
\vdots & \vdots & \vdots & \ddots & \vdots & \vdots & \vdots & \ddots & \vdots \\
\mathbf{A}_{n0} & \mathbf{A}_{n1} & \mathbf{A}_{n2} & \cdots & \mathbf{A}_{nn} & -\mathbf{B}_{n1} & -\mathbf{B}_{n2} & \cdots & -\mathbf{B}_{nn} \\
\mathbf{0} & \mathbf{A}_1^f & \mathbf{0} & \cdots & \mathbf{0} & \mathbf{B}_1^f & \mathbf{0} & \cdots & \mathbf{0} \\
\mathbf{0} & \mathbf{0} & \mathbf{A}_2^f & \cdots & \mathbf{0} & \mathbf{0} & \mathbf{B}_2^f & \cdots & \mathbf{0} \\
\vdots & \vdots & \vdots & \ddots & \vdots & \vdots & \vdots & \ddots & \vdots \\
\mathbf{0} & \mathbf{0} & \mathbf{0} & \cdots & \mathbf{A}_n^f & \mathbf{0} & \mathbf{0} & \cdots & \mathbf{B}_n^f
\end{bmatrix}
\begin{Bmatrix}
\mathbf{u}_0 \\ \mathbf{u}_1 \\ \mathbf{u}_2 \\ \vdots \\ \mathbf{u}_n \\ \mathbf{t}_1 \\ \mathbf{t}_2 \\ \vdots \\ \mathbf{t}_n
\end{Bmatrix}
$$

$$
=
\begin{bmatrix}
\mathbf{B}_{00} \\ \mathbf{B}_{10} \\ \mathbf{B}_{20} \\ \vdots \\ \mathbf{B}_{n0} \\ \mathbf{0} \\ \mathbf{0} \\ \vdots \\ \mathbf{0}
\end{bmatrix}
\{\mathbf{t}_0\},
\tag{4.88}
$$

in which $\mathbf{u}_0$ and $\mathbf{t}_0$ are the displacement and traction vector on the outer boundary S_0, $\mathbf{u}_i$ and $\mathbf{t}_i$ are the displacement and traction vector on the interface S_i from the matrix domain, $\mathbf{A}_{ij}$ and $\mathbf{B}_{ij}$ are the coefficient submatrices from the matrix domain, and $\mathbf{A}_i^f$ and $\mathbf{B}_i^f$ are the coefficient submatrices from inclusion i. By rearranging the terms in Eq. (4.88), we can write an alternative form of the BEM system of equations as [74]:

$$
\begin{aligned}
&\text{Matrix } S_0 \\
&\text{Matrix } S_1 \\
&\text{Inclusion } S_1 \\
&\text{Matrix } S_2 \\
&\text{Inclusion } S_2 \\
&\quad\vdots \\
&\text{Matrix } S_n \\
&\text{Inclusion } S_n
\end{aligned}
\begin{bmatrix}
\mathbf{A}_{00} & \mathbf{A}_{01} & -\mathbf{B}_{01} & \mathbf{A}_{02} & -\mathbf{B}_{02} & \cdots & \mathbf{A}_{0n} & -\mathbf{B}_{0n} \\
\mathbf{A}_{10} & \mathbf{A}_{11} & -\mathbf{B}_{11} & \mathbf{A}_{12} & -\mathbf{B}_{12} & \cdots & \mathbf{A}_{1n} & -\mathbf{B}_{1n} \\
\mathbf{0} & \mathbf{A}_1^f & \mathbf{B}_1^f & \mathbf{0} & \mathbf{0} & \cdots & \mathbf{0} & \mathbf{0} \\
\mathbf{A}_{20} & \mathbf{A}_{21} & -\mathbf{B}_{21} & \mathbf{A}_{22} & -\mathbf{B}_{22} & \cdots & \mathbf{A}_{2n} & -\mathbf{B}_{2n} \\
\mathbf{0} & \mathbf{0} & \mathbf{0} & \mathbf{A}_2^f & \mathbf{B}_2^f & \cdots & \mathbf{0} & \mathbf{0} \\
\vdots & \vdots & \vdots & \vdots & \vdots & \ddots & \vdots & \vdots \\
\mathbf{A}_{n0} & \mathbf{A}_{n1} & -\mathbf{B}_{n1} & \mathbf{A}_{n2} & -\mathbf{B}_{n2} & \cdots & \mathbf{A}_{nn} & -\mathbf{B}_{nn} \\
\mathbf{0} & \mathbf{0} & \mathbf{0} & \mathbf{0} & \mathbf{0} & \cdots & \mathbf{A}_n^f & \mathbf{B}_n^f
\end{bmatrix}
\begin{Bmatrix}
\mathbf{u}_0 \\ \mathbf{u}_1 \\ \mathbf{t}_1 \\ \mathbf{u}_2 \\ \mathbf{t}_2 \\ \vdots \\ \mathbf{u}_n \\ \mathbf{t}_n
\end{Bmatrix}
=
\begin{bmatrix}
\mathbf{B}_{00} \\ \mathbf{B}_{10} \\ \mathbf{0} \\ \mathbf{B}_{20} \\ \mathbf{0} \\ \vdots \\ \mathbf{B}_{n0} \\ \mathbf{0}
\end{bmatrix}
\{\mathbf{t}_0\}.
\tag{4.89}
$$

Both systems of equations in (4.88) and (4.89) can be applied with the fast multipole BEM to solve inclusion problems in elasticity. The multipole expansions and related translations discussed in the previous two sections for 2D and 3D elasticity BIEs can be applied readily for the BIEs from the matrix domain and those from the inclusion domains. The only difficult part in the implementation is the bookkeeping of the locations of the submatrices from different domains in the systems of equations.

Preconditioning for the fast multipole BEM is even more crucial for its convergence and efficiency in solving multidomain problems. Two preconditioners can be devised based on the two forms of the BEM systems of equations shown in Eqs. (4.88) and (4.89).

For *Preconditioner A*, a block diagonal preconditioner based on Eq. (4.88) is used. For the matrix domain, a diagonal submatrix is formed on each leaf by use of direct evaluations of the kernels on the elements within that leaf, whereas for the inclusions, the submatrix $\mathbf{B}_i^f$ in Eq. (4.88) along the main diagonal is used for each inclusion.

For *Preconditioner B*, a block diagonal preconditioner based on Eq. (4.89) is used. In this case, the following matrix from the matrix in Eq. (4.89) is used as the preconditioner:

$$\mathbf{M} = \begin{bmatrix} \mathbf{A}_{00} & \mathbf{0} & \mathbf{0} & \mathbf{0} & \mathbf{0} & \cdots & \mathbf{0} & \mathbf{0} \\ \mathbf{0} & \mathbf{A}_{11} & -\mathbf{B}_{11} & \mathbf{0} & \mathbf{0} & \cdots & \mathbf{0} & \mathbf{0} \\ \mathbf{0} & \mathbf{A}_1^f & \mathbf{B}_1^f & \mathbf{0} & \mathbf{0} & \cdots & \mathbf{0} & \mathbf{0} \\ \mathbf{0} & \mathbf{0} & \mathbf{0} & \mathbf{A}_{22} & -\mathbf{B}_{22} & \cdots & \mathbf{0} & \mathbf{0} \\ \mathbf{0} & \mathbf{0} & \mathbf{0} & \mathbf{A}_2^f & \mathbf{B}_2^f & \cdots & \mathbf{0} & \mathbf{0} \\ \vdots & \vdots & \vdots & \vdots & \vdots & \ddots & \vdots & \vdots \\ \mathbf{0} & \mathbf{0} & \mathbf{0} & \mathbf{0} & \mathbf{0} & \cdots & \mathbf{A}_{nn} & -\mathbf{B}_{nn} \\ \mathbf{0} & \mathbf{0} & \mathbf{0} & \mathbf{0} & \mathbf{0} & \cdots & \mathbf{A}_n^f & \mathbf{B}_n^f \end{bmatrix}. \qquad (4.90)$$

This preconditioner is equivalent to solving many inclusion problems as if there were only one inclusion embedded in an infinite domain in each case. For this preconditioner, larger diagonal matrices need to be processed for each inclusion, which can be time-consuming if the number of elements on each inclusion is large. However, this preconditioner is very effective for inclusion problems because the number of iterations for the GMRES solver can be reduced significantly, as is shown in the numerical examples in the following section (also in Ref. [74]). A similar preconditioner is applied in the 3D fast multipole BEM for modeling rigid inclusion problems by the 3D single-domain CBIE in Refs. [80, 81].

The systems in Eqs. (4.88) and (4.89) are right preconditioned with the preceding two preconditioners, respectively, in solving elasticity inclusion problems by using the fast multiple BEM. LU decompositions (LU stands for lower triangular and upper triangular matrices) of the submatrices in these preconditioners can be computed once and saved in memory in the subsequent iterations to save the CPU time in solving such problems.

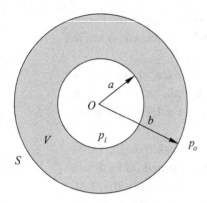

Figure 4.3. A thick cylinder with pressure loads.

4.10 Numerical Examples

Several numerical examples are given in this section to demonstrate the accuracy and efficiency of the fast multipole BEM for solving 2D and 3D elasticity problems. Most of the computations were done on a Pentium IV laptop PC with a 2.4-GHz CPU and 1-GB RAM. In all of the cases, the material has Young's modulus E and Poisson's ratio v.

4.10.1 A Cylinder with Pressure Loads

We first consider a thick cylinder under pressure loads (in the plane-strain case) as shown in Figure 4.3. The inner pressure is p_i and the outer pressure is p_o. In the case studied, $b = 2a$, $p_i = p_o = p$, and Poisson's ratio $v = 0.3$. We discretize the inner and outer boundaries with the same number of elements and run both the fast multipole BEM code and a conventional BEM code that also uses constant elements and analytical integrations. The conventional BEM code uses both the direct solver (LAPACK) and the iterative solver (GMRES) for solving the linear system. For the fast multipole BEM, the numbers of terms for both multipole and local expansions were set to 20, the maximum number of elements in a leaf to 20, and the tolerance for convergence of the solution to 10^{-6}. All the fast multipole BEM results converged in about three iterations without using any preconditioner in this example.

   Table 4.1 shows the results of radial displacement u_r and hoop stress σ_θ at the inner boundary obtained with both the fast multipole BEM and the conventional BEM (with the direct solver) as the total number of elements increases from 200 to 4800 (DOFs from 400 to 9600). As we can see, the results for both the fast multipole BEM and the conventional BEM converge quickly to the exact solution [88] for the mesh with 360 constant elements with a relative error of less than 3%. The results continue to improve with the increase of the number of elements.

Table 4.1. *Radial displacement and hoop stress at the inner boundary*

DOFs	u_r ($\times pa/E$)		σ_θ ($\times p$)	
	Conventional BEM	Fast multipole BEM	Conventional BEM	Fast multipole BEM
400	−0.52233	−0.52233	−1.00228	−1.00228
720	−0.52143	−0.52143	−1.00149	−1.00148
1440	−0.52076	−0.52076	−1.00081	−1.00082
2880	−0.52039	−0.52039	−1.00042	−1.00042
4800	−0.52024	−0.52024	−1.00026	−1.00026
9600	−0.52012	−0.52012	−1.00013	−1.00007
Exact Solution	−0.52000		−1.00000	

The CPU times used for the two BEM approaches are plotted in Figure 4.4, which shows the significant advantage of the fast multipole BEM compared with the conventional BEM with either a direct or an iterative solver. For example, for the model with 4800 elements (DOFs = 9600), the fast multipole BEM used only 3 s of the CPU time, whereas the conventional BEM used 1483 s with the direct solver and 38 s with the iterative solver. Beyond 10,000 DOFs, the conventional BEM (with double precision) encounters the 1-GB physical memory barrier and cannot run efficiently without using the virtual memory. It is also interesting to note from Figure 4.4 that the slopes of the

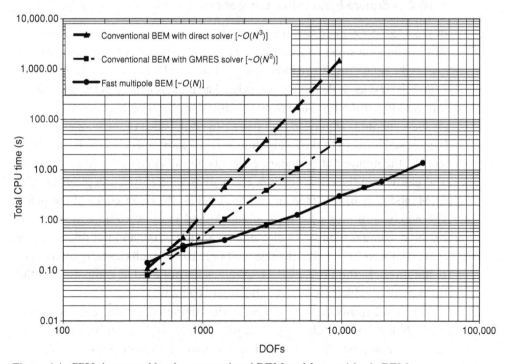

Figure 4.4. CPU times used by the conventional BEM and fast multipole BEM.

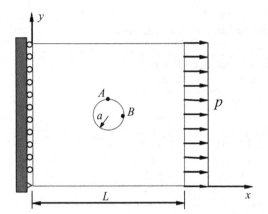

Figure 4.5. A square plate with a circular hole at the center and loaded with p.

three curves for the conventional BEM with direct solver, iterative solver, and the fast multipole BEM are close to 3, 2, and 1 on the log–log scales, suggesting $O(N^3)$, $O(N^2)$, and $O(N)$ efficiencies of the three methods, respectively.

This example shows that the fast multipole BEM is very efficient compared with the conventional BEM. In addition, the fast multipole BEM results are equally accurate as the conventional BEM results, and they are very stable with the increase of the model size.

4.10.2 A Square Plate with a Circular Hole

In the second example, we further study the accuracy of the fast multipole BEM by using a stress concentration problem – a square plate with a circular hole at the center, as shown in Figure 4.5. The edge length of the square plate is L, and the radius of the hole is $a = 0.1L$. The plate is loaded in the x direction with a uniform load p and Poisson's ratio $v = 0.3$. The maximum (at point A) and minimum (at point B) hoop stresses on the edge of the hole are sought (Figure 4.5) using both the fast multipole BEM code and ANSYS. In the BEM models, the number of boundary elements on the edge of the hole increases whereas that on the outer edges of the plate is kept at 100, except for the last BEM model, in which 200 elements are used on the outer edges of the plate. The numbers of terms for both multipole and local expansions were set to 20, the maximum number of elements in a leaf to 100, and the tolerance for convergence to 10^{-6}. All of the fast multipole BEM results converged in about 20 iterations. In the FEM models, Q4 elements are used to compare with the BEM models (which use constant boundary elements).

Table 4.2 shows the comparison of the computed hoop stresses at points A and B. For an infinitely large plate with a hole, the hoop stress at point A is $3p$ and that at point B is $-p$ [88]. For our finite-sized plate with the hole, the

Table 4.2. *Computed hoop stress σ_θ ($\times p$) on the edge of the hole*

Fast multipole BEM			FEM		
DOFs	At point A	At point B	DOFs	At point A	At point B
560	3.215	−1.176	1206	3.148	−1.101
920	3.216	−1.183	4522	3.229	−1.185
1640	3.216	−1.185	9490	3.225	−1.187
3080	3.217	−1.188	38,440	3.226	−1.192
7600	3.222	−1.190			

hoop stresses should be slightly higher than these values. The stress values for both fast multipole BEM (with DOFs = 1640) and FEM (with DOFs = 4522) converged quickly to around $3.22p$ at point A and $-1.19p$ at point B. Further increases of the numbers of elements provided little improvement in the results. This example demonstrates again that the results obtained when the fast multipole BEM code is used are accurate and stable.

It should be pointed out that the element types used for both the BEM and the FEM in this study are the simplest elements available. If higher-order elements such as quadratic elements are used, a few hundred elements should be sufficient for both the BEM and the FEM to achieve the same accuracy as reported in this example.

4.10.3 Multiple Inclusion Problems

We next study multiple inclusion problems by using the dual BIE and the fast multipole BEM [74]. The same square domain and BCs as in the previous example are used (Figure 4.5) but with elliptical inclusions (long axis a and short axis b). Two cases are considered here, one with multiple circular inclusions (long and unidirectional fibers) under the plane-strain condition and the other with multiple cracklike inclusions under plane-stress condition. For the circular inclusion case, the parameters used are $a = b = 0.2$, fiber volume fraction = 12.57%, $E = 1$, and $\nu = 0.25$ for the matrix and $E = 10$ and $\nu = 0.25$ for the inclusions. For the cracklike inclusion case, $b = 0.2$, $a/b = 0.01$, crack density = $b^2 = 4\%$, $E = 1$ and $\nu = 0.25$ for the matrix, and $E = 0.00001$ and $\nu = 0.25$ for the inclusions (cracks). In both cases, the inclusions are randomly distributed in the material domains. Two BEM models for the two cases are shown in Figure 4.6. For the outer boundary, 400 elements are used; on each interface, 200 elements are used.

We evaluate the effective Young's moduli of the materials containing the circular inclusions and cracks in the x–y plane by using the fast multipole BEM with the CHBIE and compare with the estimates by using homogenization

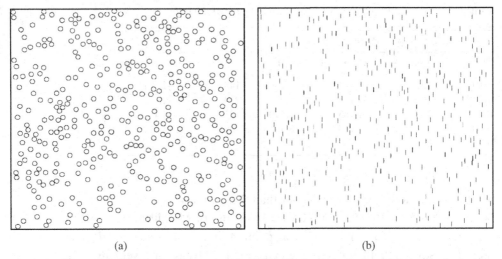

(a)   (b)

Figure 4.6. Elastic domains embedded with 400 elastic inclusions: (a) circular inclusions (fibers) with $E_i / E_0 = 10$, (b) cracklike inclusions with $E_i / E_0 = 0.00001$ and $a/b = 0.01$.

theories. Table 4.3 shows the BEM results with different numbers of the inclusions in the models, and excellent results are obtained for both cases. With the increase of the size of the models, the evaluated effective Young's moduli approach constant values, as expected.

Figure 4.7 is a plot of the CPU times used with the fast multipole BEM for the two cases studied and with the two preconditioners discussed in the previous section on the multidomain BEM. Preconditioner A is based on the coefficients calculated on leaves, and preconditioner B is based on those on the

Table 4.3. *Computed effective moduli for the materials with circular and cracklike inclusions*

No. of inclusions	Total DOFs	Effective Moduli ($\times E_0$)	
		Circular inclusions (E_x, E_y)	Cracklike inclusions (E_x)
2×2	4000	1.2678	0.7631
4×4	13,600	1.2728	0.8024
6×6	29,600	1.2596	0.7808
8×8	52,000	1.2605	0.7923
10×10	80,800	1.2649	0.7891
12×12	116,000	1.2640	0.7902
14×14	157,600	1.2635	0.7886
16×16	205,600	1.2651	0.7900
18×18	260,000	1.2642	0.7897
20×20	320,800	1.2644	0.7885
Analytical estimates [89]		1.2491	0.7992

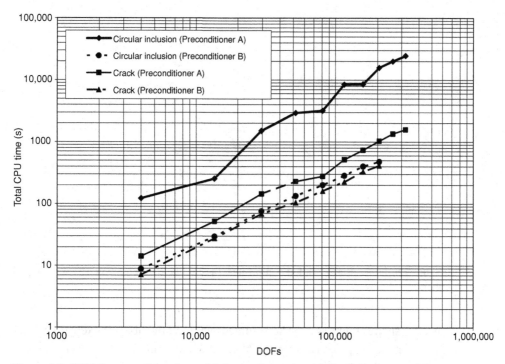

Figure 4.7. CPU times used for the multiple circular and cracklike inclusion models.

inclusions [74]. The computer used for these calculations is a desktop PC with an Intel Pentium D 3.2-GHz processor and 2-GB memory size. For the circular inclusion case, the numbers of iterations using preconditioner A range from 141 to 550, whereas those using preconditioner B range from 11 to 16, with a tolerance of 10^{-6}. For the cracklike inclusion case, the numbers of iterations using preconditioner A range from 29 to 41, whereas those using preconditioner B range from 12 to 14. Significant advantages of using preconditioner B are observed in both cases. It is also observed that the dual BIE (CHBIE) formulation, which uses a linear combination of the CBIE and HBIE, provides much better conditioning for the cracklike inclusion problems, based on the fact that much faster convergence was achieved for the cracklike problem than for the circular inclusion problem.

4.10.4 Modeling of Functionally Graded Materials

We next show an example in modeling functionally graded materials by using the fast multipole BEM for 2D elasticity inclusion problems. We model functionally graded composites with long fibers that are distributed randomly and with a decreasing density in one direction. The plane-strain model is used to extract the moduli of the composites in the transverse directions. Two cases are studies, one with the fibers of a circular shape and the other with the fibers

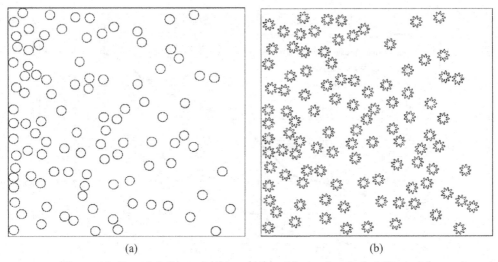

Figure 4.8. Functionally graded composites: (a) circular-shaped fibers; (b) star-shaped fibers.

of a star shape (Figure 4.8). The volume fraction of the fibers is fixed at 12.57%, and the material properties for the fibers and matrix are the same as in the previous example [Figure 4.6(a)].

Table 4.4 shows the computed effective Young's moduli for the functionally graded composite models. The cell sizes increase with the increase of the included fibers to keep the fiber volume fraction at 12.57%, as in the previous example. It turns out that the effective moduli converge to the same value of the analytical solution [89] that is based on a "uniform" (nongraded) distribution of the fibers as used in the previous example [Figure 4.6(a)]. This suggests

Table 4.4. *Computed effective moduli for the functionally graded composites*

No. of inclusions	Total DOFs	Effective Moduli E_x ($\times E_0$)	
		Circular-shaped fibers	Star-shaped fibers
4	4000	1.2752	1.2254
16	13,600	1.2600	1.2181
36	29,600	1.2674	1.2227
64	52,000	1.2645	1.2208
100	80,800	1.2639	1.2227
144	116,000	1.2616	1.2200
196	157,600	1.2621	1.2216
256	205,600	1.2600	1.2209
324	260,000	1.2625	1.2224
400	320,800	1.2628	1.2215
Analytical estimates [89]		1.2491	

that the effective moduli of the composites are not affected by the shape and the distribution of the fibers, as expected, if the volume fraction of the fibers is the same. However, the graded distribution of the fibers may reduce the stresses in the composites, and the star-shaped fibers may improve the interfacial properties of the composites.

4.10.5 Large-Scale Modeling of Fiber-Reinforced Composites

The 3D BEM was applied to model fiber-reinforced composite materials for quite some time (see, e.g., Refs. [54, 90]). However, because of the limitations in the conventional BEM, only models with a few fibers can be modeled and analyzed with the BEM, even with parallel computing techniques. With the advance of the fast multipole BEM, we can now analyze 3D BEM models of composite materials with more than tens of thousands of fibers that are modeled explicitly.

To demonstrate this, we next study a few 3D representative volume elements (RVEs) of fiber-reinforced composites by using the fast multipole BEM for 3D problems. In these models, the fibers are regarded as rigid inclusions and the matrix material is considered as an elastic medium. These models are useful in cases in which the Young's moduli of the fibers are much higher than those of the matrix, as in the case of carbon nanotube (CNT) composites. The RVE is loaded in the x direction to evaluate the effective modulus in that direction. Elements are needed on only the boundaries and interfaces with the BEM. More examples in modeling fiber-reinforced composites, especially CNT composites, including those incorporating the molecular dynamics (MD) and cohesive interface conditions, can be found in Refs. [80–82]. The following large models were solved on a supercomputer at Kyoto University.

Figure 4.9 shows a smaller RVE of a short fiber-reinforced composite and the stress contour plot on the interfaces between the fibers and matrix. A boundary element mesh using constant triangular elements on the interfaces is shown in the insert. Using the fast multipole BEM, we can readily study the interface stresses in such models and extract the effective mechanical properties of the composites.

Figure 4.10 shows a larger RVE model with 5832 long fibers and 10,532,592 DOFs. The volume fraction of the fibers is 3.85%, and Poisson's ratio for the matrix is 0.3. Figure 4.11 shows the normalized effective Young's moduli (E_{eff}/E_{matrix}) computed from the RVE models for the composite with increasing numbers of the fibers (from 9 to 5832) and with uniform and random distributions of the fibers, while keeping the same value of the volume fraction. Theoretically, the moduli of the materials should be independent of the sizes of the RVE models. However, when the RVE sizes are small (with only a few fibers), there are significant changes in the estimated moduli as we increase the model sizes, which suggests that the models are not yet representative

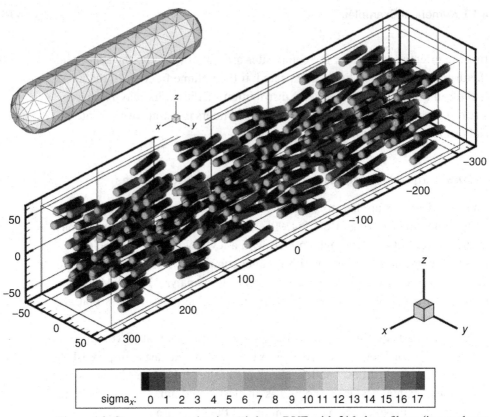

Figure 4.9. Stress contour plot ($\times \sigma^{\infty}$) for a RVE with 216 short fibers (insert shows the mesh).

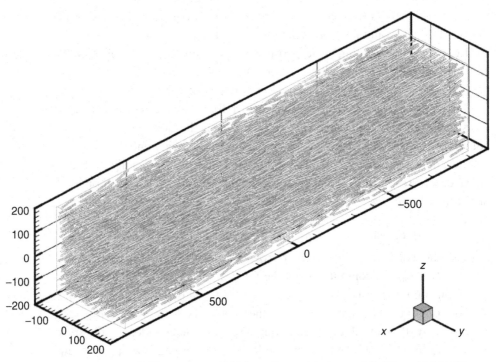

Figure 4.10. A RVE containing 5832 long fibers with the total DOFs = 10,532,592.

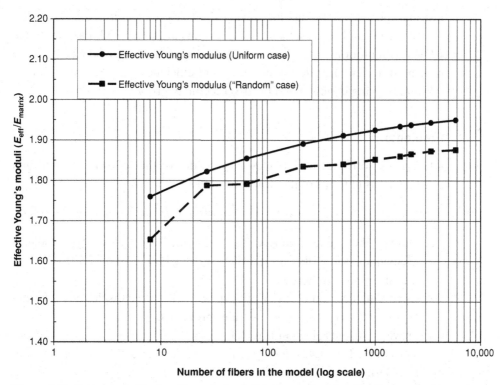

Figure 4.11. Estimated effective Young's moduli in the x direction for the composite models.

of the composites. The estimated moduli approach constant with the increase of the fibers and thus the sizes of the models. The increases in the values of the moduli compared with those of the matrix range from 75.9% to 95.0% for the uniform case and from 65.4% to 87.6% for the random case. However, the increases in the random case are about 8% lower than those in the uniform case. This suggests that even small misalignments and rotations of fibers can offset the enhancement in the stiffness of composites.

The largest BEM model solved so far in modeling fiber-reinforced composites contains 16,000 fibers with a total of 28.8 million DOFs and was solved on the supercomputer at Kyoto University [80]. More discussion and examples of modeling the fiber-reinforced composites using the fast multipole BEM can be found in Refs. [80–82].

4.11 Summary

In this chapter, the governing equations for elastostatic problems are reviewed. The fundamental solutions, their properties, and the generalized Green's identity (Somigliana's identity) are introduced. The BIE formulations are presented based on this identity and the fundamental solutions. The fast multipole formulations for solving the BIEs are described in detail for 2D problems, and the results for 3D problems are presented. Fast multipole BEMs for multidomain problems are also discussed. Numerical examples are presented to demonstrate the accuracy and efficiencies of the fast multipole BEM for solving large-scale 2D and 3D elasticity problems.

Problems

4.1. For an isotropic, linearly elastic body, derive the following equilibrium equations in terms of the displacement field u_i:

$$\mu u_{i,jj} + \frac{\mu}{1-2\nu}u_{j,ji} + f_i = 0. \tag{4.91}$$

4.2. Derive the generalized Green's identity for elasticity problems (Somigliana's identity) in Eq. (4.17) by using both the Gauss theorem and the virtual work theorem.

4.3. Derive the traction kernel T_{ij} in Eq. (4.10) from the displacement kernel U_{ij} in Eq. (4.9) for 2D (plane-strain) problems.

4.4. Prove the results given in Eq. (4.32).

4.5. Verify that Eq. (4.42) does give the same fundamental solution given in Eq. (4.9).

4.6. Prove the complex notation of the traction given in Eq. (4.48).

4.7. Write a conventional BEM code for solving 2D elasticity problems using constant elements. Use the 2D potential code used in Chapter 2 (Appendix B.1) as the starting point and use the analytical integration results in Appendix A.2 for the 2D elasticity case.

4.8. Solve the cylinder model in Figure 4.3 by using a quarter-symmetry model and compare the accuracy and efficiency of the results obtained with those using the full model.

4.9. Write a fast multipole BEM code for solving 2D elasticity problems using constant elements based on the 2D potential fast multipole BEM code given in Appendix B.2. Study its accuracy and efficiency by using the cylinder example.

5 Stokes Flow Problems

Stokes flows are incompressible flows at low Reynolds' number [91], which can be found in many applications such as creeping flows in biological systems and fluid–structure interactions in MEMSs. Stokes flow problems were formulated with BIEs and solved by the BEM for decades with either direct or indirect BIE formulations (see, e.g., Refs. [92, 93]).

For Stokes flow problems using the fast multipole BEM, there are several approaches reported in the literature. Greengard *et al.* [68] developed a fast multipole formulation for directly solving the biharmonic equations in 2D elasticity with the Stokes flow as a special case. Gomez and Power [37] studied 2D cavity flow governed by Stokes equations by using both direct and indirect BIEs and the FMM in which they used Taylor series expansions of the kernels in real variables directly. Mammoli and Ingber [40] applied the fast multipole BEM to study Stokes flow around cylinders in a bounded 2D domain by using direct and indirect BIEs with the kernels expanded by a Taylor series of the real variables. In the context of modeling a MEMS, Ding and Ye [94] developed a fast BEM by using the precorrected fast Fourier transform (FFT) accelerated technique for computing drag forces with 3D MEMS models with slip BCs. Frangi and co-workers [95–98] conducted extensive research by using the direct BIE formulations and the fast multipole BEM for evaluating damping forces of 3D MEMS structures. They applied a mixed-velocity–traction BIE in modeling large-scale 3D MEMS problems under both no-slip and slip BCs. Liu [99] also developed a fast multipole BEM for solving 2D Stokes flow problems based on a dual BIE formulation.

In this chapter, the direct BIE formulations for solving Stokes flow problems are reviewed first. Then, the FMMs for both 2D and 3D Stokes flow problems are presented. Because of the similarities of the Stokes flow equations to those for the elasticity problems, many of the results for Stokes flow problems can be obtained directly from those for elasticity presented in the previous chapter. Several numerical examples are presented, which clearly show the accuracy, efficiency, and potentials of the fast multipole BEM for analyzing large-scale Stokes flow problems.

5.1 The Boundary-Value Problem

Consider the following boundary-value problem for a steady-state Stokes flow in domain V with boundary S:

Equilibrium equations:

$$\sigma_{ij,j} = 0, \quad \text{in } V; \tag{5.1}$$

Continuity equations:

$$u_{i,i} = 0, \quad \text{in } V; \tag{5.2}$$

Constitutive equations:

$$\sigma_{ij} = -p\delta_{ij} + \mu(u_{i,j} + u_{j,i}), \quad \text{in } V, \tag{5.3}$$

where σ_{ij} is the stress in the fluid, u_i is the velocity, p is the pressure, and μ is the coefficient of viscosity of the fluid. Substituting Eqs. (5.2) and (5.3) into Eq. (5.1), we obtain the *Stokes equation* as follows:

$$-p_{,i} + \mu u_{i,jj} = 0, \quad \text{in } V. \tag{5.4}$$

Taking the derivative of this equation again and applying Eq. (5.2), we note that the pressure p field satisfies the Laplace equation:

$$\nabla^2 p = 0, \quad \text{in } V. \tag{5.5}$$

Two typical BCs for Stokes flow problems are:

$$u_i = \overline{u_i} \quad \text{on } S_u \text{ (velocity BCs)}, \tag{5.6}$$

$$t_i = \sigma_{ij}n_j = \overline{t_i} \quad \text{on } S_t \text{ (traction BCs)}, \tag{5.7}$$

where the overbar indicates the specified value of the field, t_i is the traction, n_i are the components of the outward normal n, and $S_u \cup S_t = S$.

5.2 Fundamental Solution for Stokes Flow Problems

Let $\Sigma_{ijk}(\mathbf{x}, \mathbf{y})$, $U_{ij}(\mathbf{x}, \mathbf{y})$, and $P_i(\mathbf{x}, \mathbf{y})$ be the stress, velocity, and pressure fields, respectively, in the fundamental solution for Stokes flow problems, with i indicating the direction of the unit concentrate force acting at the source point $\mathbf{x}$. We have the following equations:

$$\Sigma_{ijk,k}(\mathbf{x}, \mathbf{y}) + \delta_{ij}\delta(\mathbf{x}, \mathbf{y}) = 0, \quad \forall \mathbf{x}, \mathbf{y} \in R^2/R^3, \tag{5.8}$$

$$U_{ij,j}(\mathbf{x}, \mathbf{y}) = 0, \quad \forall \mathbf{x}, \mathbf{y} \in R^2/R^3, \tag{5.9}$$

in which the Dirac δ function $\delta(\mathbf{x}, \mathbf{y})$ represents the body force corresponding to the unit force. From Eq. (5.3), we have:

$$\Sigma_{ijk} = -P_i \delta_{jk} + \mu(U_{ij,k} + U_{ik,j}). \tag{5.10}$$

Substituting this result into Eq. (5.8) and applying Eq. (5.9), we have:

$$-P_{i,j}(\mathbf{x}, \mathbf{y}) + \mu U_{ij,kk}(\mathbf{x}, \mathbf{y}) + \delta_{ij}\delta(\mathbf{x}, \mathbf{y}) = 0, \quad \forall \mathbf{x}, \mathbf{y} \in R^2/R^3. \tag{5.11}$$

Taking the derivative again with respect to y_j, we obtain:

$$-\nabla^2 P_i(\mathbf{x}, \mathbf{y}) + \frac{\partial}{\partial y_i}\delta(\mathbf{x}, \mathbf{y}) = 0, \quad \forall \mathbf{x}, \mathbf{y} \in R^2/R^3. \tag{5.12}$$

Comparing this equation with Eq. (2.4), we find that:

$$P_i(\mathbf{x}, \mathbf{y}) = -G_{,i}(\mathbf{x}, \mathbf{y}) = -\frac{\partial G(\mathbf{x}, \mathbf{y})}{\partial y_i}, \tag{5.13}$$

where G is the fundamental solution for potential problems.

For 2D Stokes flow problems, the pressure, velocity, and traction fields in the fundamental solution are given by:

$$P_i(\mathbf{x}, \mathbf{y}) = \frac{1}{2\pi r}r_{,i}, \tag{5.14}$$

$$U_{ij}(\mathbf{x}, \mathbf{y}) = \frac{1}{4\pi\mu}\left[\delta_{ij}\log\left(\frac{1}{r}\right) + r_{,i}r_{,j} - \frac{1}{2}\delta_{ij}\right], \tag{5.15}$$

$$T_{ij}(\mathbf{x}, \mathbf{y}) = -\frac{1}{\pi r}r_{,i}r_{,j}r_{,k}n_k(\mathbf{y}). \tag{5.16}$$

For 3D Stokes flow problems, the fundamental solution gives:

$$P_i(\mathbf{x}, \mathbf{y}) = \frac{1}{4\pi r^2}r_{,i}, \tag{5.17}$$

$$U_{ij}(\mathbf{x}, \mathbf{y}) = \frac{1}{8\pi\mu r}(\delta_{ij} + r_{,i}r_{,j}), \tag{5.18}$$

$$T_{ij}(\mathbf{x}, \mathbf{y}) = -\frac{3}{4\pi r^2}r_{,i}r_{,j}r_{,k}n_k(\mathbf{y}). \tag{5.19}$$

It is noticed that the U and T kernels for Stokes flow problems also can be obtained readily from the U and T kernels for elasticity problems, respectively, by simply setting Poisson's ratio $\nu = \frac{1}{2}$ in the results for elasticity problems. Similar *integral identities* as those in Eqs. (4.13)–(4.16) are satisfied by the fundamental solutions for Stokes flow problems.

5.3 Boundary Integral Equation Formulations

Applying a generalized Green's identity, similar to the one in Eq. (4.17), we obtain the following representation integral for the velocity within the domain V:

$$u_i(\mathbf{x}) = \int_S [U_{ij}(\mathbf{x}, \mathbf{y})t_j(\mathbf{y}) - T_{ij}(\mathbf{x}, \mathbf{y})u_j(\mathbf{y})]dS(\mathbf{y}), \quad \forall \mathbf{x} \in V. \tag{5.20}$$

Let the source point $\mathbf{x}$ approach the boundary S; we obtain the CBIE for Stokes flow problems (see, e.g., Refs. [92, 93]):

$$c_{ij}(\mathbf{x})u_j(\mathbf{x}) = \int_S [U_{ij}(\mathbf{x}, \mathbf{y})t_j(\mathbf{y}) - T_{ij}(\mathbf{x}, \mathbf{y})u_j(\mathbf{y})]dS(\mathbf{y}), \quad \forall \mathbf{x} \in S, \qquad (5.21)$$

where $c_{ij} = \frac{1}{2}\delta_{ij}$ if S is smooth around $\mathbf{x}$, the integral with the U kernel is a weakly singular integral, and the integral with the T kernel is a Cauchy principal-value integral. Equation (5.21) is valid for both interior and exterior problems (assuming velocity and traction fields vanish at the infinity for the latter). Equation (5.21) is the *direct* BIE formulation for Stokes flow problems in which the density functions have direct physical meanings; that is, they represent the velocity and traction (u_i and t_i, respectively).

The pressure field can be represented by the following integral (see also Refs. [92, 93]):

$$p(\mathbf{x}) = \int_S [G_{,j}(\mathbf{x}, \mathbf{y})t_j(\mathbf{y}) - 2\mu F_{,j}(\mathbf{x}, \mathbf{y})u_j(\mathbf{y})]dS(\mathbf{y}), \quad \forall \mathbf{x} \in V, \qquad (5.22)$$

in which $F(\mathbf{x}, \mathbf{y})$ is the same F kernel for potential problems; that is:

$$F(\mathbf{x}, \mathbf{y}) = \frac{\partial G(\mathbf{x}, \mathbf{y})}{\partial n(\mathbf{y})}. \qquad (5.23)$$

From Eq. (5.22), we can find the pressure field $p(\mathbf{x})$ in domain V once the velocity and traction fields are known on boundary S.

Taking the derivatives of Eq. (5.20) and applying Eq. (5.3), we have the following results:

$$\sigma_{ij}(\mathbf{x})n_j(\mathbf{x}) = -p(\mathbf{x})n_i(\mathbf{x}) + n_i(\mathbf{x}) \int_S [G_{,j}(\mathbf{x}, \mathbf{y})t_j(\mathbf{y}) - 2\mu F_{,j}(\mathbf{x}, \mathbf{y})u_j(\mathbf{y})] \, dS(\mathbf{y})$$
$$+ \int_S [K_{ij}(\mathbf{x}, \mathbf{y})t_j(\mathbf{y}) - H_{ij}(\mathbf{x}, \mathbf{y})u_j(\mathbf{y})] \, dS(\mathbf{y}), \quad \forall \mathbf{x} \in V, \qquad (5.24)$$

with $n_i(\mathbf{x})$ being a vector at $\mathbf{x}$. Noting Eq. (5.22) and letting $\mathbf{x}$ tend to S, we obtain the following traction BIE (HBIE) from the preceding result:

$$\tilde{c}_{ij}(\mathbf{x})t_j(\mathbf{x}) = \int_S [K_{ij}(\mathbf{x}, \mathbf{y})t_j(\mathbf{y}) - H_{ij}(\mathbf{x}, \mathbf{y})u_j(\mathbf{y})] \, dS(\mathbf{y}), \quad \forall \mathbf{x} \in S, \qquad (5.25)$$

where $\tilde{c}_{ij} = \frac{1}{2}\delta_{ij}$, assuming S is smooth around $\mathbf{x}$. For 2D Stokes flow problems, the two new kernels are:

$$K_{ij}(\mathbf{x}, \mathbf{y}) = \frac{1}{\pi r} r_{,i}\, r_{,j}\, r_{,k}\, n_k(\mathbf{x}), \qquad (5.26)$$

$$H_{ij}(\mathbf{x}, \mathbf{y}) = \frac{\mu}{\pi r^2} \left[(\delta_{ij}r_{,k} + \delta_{jk}r_{,i} - 8r_{,i}\, r_{,j}\, r_{,k})\, r_{,l}\, n_l(\mathbf{y}) \right.$$
$$\left. + n_i r_{,j}\, r_{,k} + n_k r_{,i}\, r_{,j} + \delta_{ik} n_j \right] n_k(\mathbf{x}), \qquad (5.27)$$

where $n_i(\mathbf{x})$ is the normal at the source point $\mathbf{x}$. For 3D problems, the two new kernels are:

$$K_{ij}(\mathbf{x}, \mathbf{y}) = \frac{3}{4\pi r^2} r_{,i} r_{,j} r_{,k} n_k(\mathbf{x}), \tag{5.28}$$

$$H_{ij}(\mathbf{x}, \mathbf{y}) = \frac{\mu}{4\pi r^3} \left\{ [3(\delta_{ij} r_{,k} + \delta_{jk} r_{,i}) - 30 r_{,i} r_{,j} r_{,k}] r_{,l} n_l(\mathbf{y}) \right.$$
$$\left. + 3(n_i r_{,j} r_{,k} + n_k r_{,i} r_{,j}) + 2\delta_{ik} n_j \right\} n_k(\mathbf{x}). \tag{5.29}$$

In HBIE (5.25), the integral with the K kernel is a CPV integral, whereas the one with the H kernel is a HFP integral. For exterior problems, it has been assumed that the pressure field $p(\mathbf{x})$ vanishes at infinity in the derivation of HBIE (5.25).

We can obtain CBIE (5.21) and HBIE (5.25) with the four kernels U_{ij}, T_{ij}, K_{ij}, and H_{ij} from those for elasticity problems by simply setting Poisson's ratio to $\frac{1}{2}$ in the corresponding elasticity BIEs. However, it is still beneficial to derive these BIEs based on the field equations in order to better understand the BIEs for Stokes flow problems. In addition, it should be pointed out that the relations between the elasticity BIEs and the Stokes flow BIEs do not provide an easy path for solving Stokes flow problems by just using the fast multipole BEM code for elasticity problems. A few results related to the elasticity BIEs become invalid when Poisson's ratio ν is set to $\frac{1}{2}$ directly (e.g., the Lamé constant $\lambda = 2\mu\nu/(1 - 2\nu) \to \infty$ when $\nu \to \frac{1}{2}$).

Some observations on CBIE (5.21) and HBIE (5.25) are in order:

1. For *a Dirichlet problem* in which velocity is prescribed on the entire boundary S, CBIE (5.21) is reduced to:

$$\int_S U_{ij}(\mathbf{x}, \mathbf{y}) t_j(\mathbf{y}) dS(\mathbf{y}) = b_i(\mathbf{x}), \quad \forall \mathbf{x} \in S, \tag{5.30}$$

where b_i is a known vector from the velocity field; HBIE (5.25) is reduced to:

$$\frac{1}{2} t_i(\mathbf{x}) = \int_S K_{ij}(\mathbf{x}, \mathbf{y}) t_j(\mathbf{y}) dS(\mathbf{y}) + d_i(\mathbf{x}), \quad \forall \mathbf{x} \in S, \tag{5.31}$$

where d_i is another known vector (assuming S is smooth). Equation (5.30), a Helmholtz equation of the first kind, is often ill-conditioned and not suitable for iterative solvers, whereas Eq. (5.31), a Helmholtz equation of the second kind, often yields a system of equations with better conditioning [37, 40, 92, 93].

2. Any constant-pressure field $p(\mathbf{x}) = p_0$, with $u_i = 0$ and $t_i = -p_0 n_i$, is a solution of both Eq. (5.30) (for *interior* and *exterior* problems) and Eq. (5.31) (for *interior* problems only). That is, $t_i = -p_0 n_i$ are eigenfunctions of both Eq. (5.30) and Eq. (5.31), although corresponding to different

eigenvalues, and their solutions for the traction field may not be unique [37, 40, 92, 93].

3. HBIE (5.25) has another "defect"; that is, an arbitrary constant can be added to the velocity field on a closed contour without changing HBIE (5.25) because:

$$\int_{S_k} H_{ij}(\mathbf{x}, \mathbf{y}) dS(\mathbf{y}) = 0,$$

for any closed contour S_k [46]. This means that we have either nonunique solutions of the velocity on the contour if traction is prescribed or inaccurate evaluation of this contour integral if velocity is given, when HBIE (5.25) is applied alone. This deficiency with the HBIE and its remedies have been discussed in the context of elasticity in Refs. [100, 101].

A remedy to the previously mentioned defects or difficulties is to use CBIE (5.21) and HBIE (5.25) together in the form of a linear combination, which was found to be very effective for 3D exterior Stokes flow problems in Refs. [95–98] and for both 2D and 3D interior and exterior potential and elasticity problems as in previous chapters. Other remedies include the so-called completed indirect BIE formulations [37, 40, 92, 93], which have been shown to yield BEM equations with better conditioning for solving Stokes flow problems.

In operator or matrix form, CBIE (5.21) and HBIE (5.25) can be written as:

$$\frac{1}{2}\mathbf{u} + \mathbf{Tu} = \mathbf{Ut}, \quad -\frac{1}{2}\mathbf{t} + \mathbf{Kt} = \mathbf{Hu},$$

respectively. Thus, a dual BIE formulation using a linear combination of CBIE (5.21) and HBIE (5.25) can be written as:

$$\left(\frac{1}{2}\mathbf{u} + \mathbf{Tu} - \mathbf{Ut}\right) + \beta\left(-\frac{1}{2}\mathbf{t} + \mathbf{Kt} - \mathbf{Hu}\right) = \mathbf{0}, \tag{5.32}$$

where β is the coupling constant. A positive β (e.g., $\beta = 1$) was found to work quite well for all the cases studied for 2D Stokes flow problems. More discussion on the selections of β can be found in Refs. [50, 52, 57, 59, 85, 102] for other problems. As mentioned in the previous chapters, dual BIE formulations are especially beneficial to the fast multipole BEM because they can provide better conditioning for the BEM systems of equations and thus can facilitate faster convergence when the iterative solvers are used with the fast multipole BEM.

5.4 Fast Multipole Boundary Element Method for 2D Stokes Flow Problems

The fast multipole algorithms for solving 2D potential and elasticity problems were described in detail in the previous two chapters. As a case similar to 2D elasticity, the 2D Stokes flow case can be handled by using the same algorithms as in 2D elasticity. The only task is to derive the required expansions and moments. For both CBIE (5.21) and HBIE (5.25), the results can be extracted from those for the 2D elasticity case given in the previous chapter. Therefore, only the results for Stokes flow problems without detailed derivations are listed.

In the previous chapter, it is shown that the two integrals in the CBIE for 2D elasticity can be represented readily in complex variables if the fundamental solutions $U_{ij}(\mathbf{x}, \mathbf{y})$ and $T_{ij}(\mathbf{x}, \mathbf{y})$ are written in complex form by using the results in 2D elasticity. By setting Poisson's ratio to $\frac{1}{2}$ in these results, we obtain the corresponding expressions for 2D Stokes flow problems. For example, the first integral in CBIE (5.21) can be written in the following complex form [cf. Eq. (4.44) for 2D elasticity]:

$$
\begin{aligned}
D_t(z_0) \equiv [\Delta_1(\mathbf{x}) + i\Delta_2(\mathbf{x})]_t &\equiv \left[\int_S U_{1j}(\mathbf{x}, \mathbf{y})t_j(\mathbf{y})dS(\mathbf{y}) \right] \\
&+ i\left[\int_S U_{2j}(\mathbf{x}, \mathbf{y})t_j(\mathbf{y})dS(\mathbf{y}) \right] \\
&= \frac{1}{4\mu} \int_S \left[G(z_0, z)t(z) + \overline{G(z_0, z)}t(z) - (z_0 - z)\overline{G'(z_0, z)t(z)} \right] dS(z),
\end{aligned}
$$

(5.33)

where $i = \sqrt{-1}$; the overbar indicates the complex conjugate; $t = t_1 + it_2$ is the complex traction; $z_0 (= x_1 + ix_2)$ and $z(= y_1 + iy_2)$ represent $\mathbf{x}$ and $\mathbf{y}$, respectively; $G(z_0, z) = -(1/2\pi)\log(z_0 - z)$ is the complex Green's function for 2D potential problems; and $G'(z_0, z) \equiv \partial G/\partial z_0$. The integral in Eq. (5.33) can be used to evaluate readily the U kernel integral in CBIE (5.21).

Similarly, the complex representation for the second integral with the T kernel in CBIE (5.21) can be written as follows [cf. Eq. (4.45) for 2D elasticity]:

$$
\begin{aligned}
D_u(z_0) \equiv [\Delta_1(\mathbf{x}) + i\Delta_2(\mathbf{x})]_u &\equiv \left[\int_S T_{1j}(\mathbf{x}, \mathbf{y})u_j(\mathbf{y})dS(\mathbf{y}) \right] \\
&+ i\left[\int_S T_{2j}(\mathbf{x}, \mathbf{y})u_j(\mathbf{y})dS(\mathbf{y}) \right] \\
&= -\frac{1}{2} \int_S \left\{ G'(z_0, z)n(z)u(z) - (z_0 - z)\overline{G''(z_0, z)n(z)u(z)} \right. \\
&\left. + \overline{G'(z_0, z)}\left[n(z)\overline{u(z)} + \overline{n(z)}u(z) \right] \right\} dS(z),
\end{aligned}
$$

(5.34)

in which $u = u_1 + iu_2$ and $n = n_1 + in_2$ are the complex velocity and normal, respectively.

The first integral with the K kernel in HBIE (5.25) can be written in the following complex form [cf. Eq. (4.52) for 2D elasticity]:

$$
\begin{aligned}
F_t(z_0) &\equiv [F_1(\mathbf{x}) + i F_2(\mathbf{x})]_t \equiv \left[\int_S K_{1j}(\mathbf{x}, \mathbf{y}) t_j(\mathbf{y}) dS(\mathbf{y}) \right] \\
&\quad + i \left[\int_S K_{2j}(\mathbf{x}, \mathbf{y}) t_j(\mathbf{y}) dS(\mathbf{y}) \right] \\
&= \frac{1}{2} \int_S \{ [G'(z_0, z) t(z) + \overline{G'(z_0, z) t(z)}] n(z_0) \\
&\quad + [\overline{G'(z_0, z)} t(z) - (z_0 - z) \overline{G''(z_0, z) t(z)}] n(z_0) \} dS(z).
\end{aligned} \tag{5.35}
$$

Similarly, the second integral with the H kernel in HBIE (5.25) can be written as follows [cf. Eq. (4.53) for 2D elasticity]:

$$
\begin{aligned}
F_u(z_0) &\equiv [F_1(\mathbf{x}) + i F_2(\mathbf{x})]_u \equiv \left[\int_S H_{1j}(\mathbf{x}, \mathbf{y}) u_j(\mathbf{y}) dS(\mathbf{y}) \right] \\
&\quad + i \left[\int_S H_{2j}(\mathbf{x}, \mathbf{y}) u_j(\mathbf{y}) dS(\mathbf{y}) \right] \\
&= -\mu \int_S \left([G''(z_0, z) n(z) u(z) + \overline{G''(z_0, z) n(z) u(z)}] n(z_0) \right. \\
&\quad + \{ \overline{G''(z_0, z)} [n(z) \overline{u(z)} + \overline{n(z)} u(z)] \\
&\quad \left. - (z_0 - z) \overline{G'''(z_0, z) n(z) u(z)} \} n(z_0) \right) dS(z).
\end{aligned} \tag{5.36}
$$

In the following subsections, the multipole expansions, local expansions, and their translations related to Eqs. (5.33) and (5.34) in the fast multipole BEM for CBIE (5.21) are presented. Then, we discuss those related to Eqs. (5.35) and (5.36) for HBIE (5.25).

5.4.1 Multipole Expansion (Moments) for the U Kernel Integral

It can be shown that the *multipole expansion* for the integral $D_t(z_0)$ is as follows [cf. Eq. (4.55) for 2D elasticity]:

$$
\begin{aligned}
D_t(z_0) &= \frac{1}{8\pi\mu} \left[\sum_{k=0}^{\infty} O_k(z_0 - z_c) M_k(z_c) + z_0 \sum_{k=0}^{\infty} \overline{O_{k+1}(z_0 - z_c) M_k(z_c)} \right. \\
&\quad \left. + \sum_{k=0}^{\infty} \overline{O_k(z_0 - z_c) N_k(z_c)} \right],
\end{aligned} \tag{5.37}
$$

where z_c is the expansion point close to the field point z (see Figure 3.2), and the two sets of *moments* about z_c are:

$$M_k(z_c) = \int_{S_c} I_k(z - z_c)t(z)dS(z), \quad \text{for } k \geq 0, \tag{5.38}$$

$$N_0 = \int_{S_c} t(z)dS(z); \tag{5.39}$$

$$N_k(z_c) = \int_{S_c} \left[\overline{I_k(z - z_c)}t(z) - \overline{I_{k-1}(z - z_c)}z\overline{t(z)} \right]dS(z), \quad \text{for } k \geq 1.$$

5.4.2 Moment-to-Moment Translation

If point z_c is moved to a new location $z_{c'}$ (see Figure 3.2), we have the following M2M translations:

$$M_k(z_{c'}) = \sum_{l=0}^{k} I_{k-l}(z_c - z_{c'})M_l(z_c), \quad \text{for } k \geq 0, \tag{5.40}$$

$$N_k(z_{c'}) = \sum_{l=0}^{k} \overline{I_{k-l}(z_c - z_{c'})}N_l(z_c), \quad \text{for } k \geq 0, \tag{5.41}$$

which are exactly the same as those used in the 2D elasticity case.

5.4.3 Local Expansion and Moment-to-Local Translation

We have the following *local expansion* [cf. Eq. (4.60) for 2D elasticity]:

$$D_t(z_0) = \frac{1}{8\pi\mu} \left[\sum_{l=0}^{\infty} L_l(z_L)I_l(z_0 - z_L) - z_0 \sum_{l=1}^{\infty} \overline{L_l(z_L)}I_{l-1}(z_0 - z_L) \right.$$

$$\left. + \sum_{l=0}^{\infty} K_l(z_L)\overline{I_l(z_0 - z_L)} \right], \tag{5.42}$$

where z_L is the local expansion point close to point z_0 (see Figure 3.2), and the coefficients are given by the following M2L translations:

$$L_l(z_L) = (-1)^l \sum_{k=0}^{\infty} O_{l+k}(z_L - z_c)M_k(z_c), \quad \text{for } l \geq 0, \tag{5.43}$$

$$K_l(z_L) = (-1)^l \sum_{k=0}^{\infty} \overline{O_{l+k}(z_L - z_c)}N_k(z_c), \quad \text{for } l \geq 0, \tag{5.44}$$

which are the same as those used in the 2D elasticity case.

5.4.4 Local-to-Local Translation

If the point for the local expansion is moved from z_L to $z_{L'}$ (see Figure 3.2), the new local expansion coefficients are given by the following L2L translations:

$$L_l(z_{L'}) = \sum_{m=l}^{\infty} I_{m-l}(z_{L'} - z_L) L_m(z_L), \quad \text{for } l \geq 0, \tag{5.45}$$

$$K_l(z_{L'}) = \sum_{m=l}^{\infty} \overline{I_{m-l}(z_{L'} - z_L)} K_m(z_L), \quad \text{for } l \geq 0, \tag{5.46}$$

which are also the same as those used in the 2D elasticity case.

5.4.5 Expansions for the T Kernel Integral

The multipole expansion of (5.34) can be written as follows [cf. Eq. (4.65) for 2D elasticity]:

$$D_u(z_0) = \frac{1}{4\pi} \left[\sum_{k=1}^{\infty} O_k(z_0 - z_c) \widetilde{M}_k(z_c) + z_0 \sum_{k=1}^{\infty} \overline{O_{k+1}(z_0 - z_c) \widetilde{M}_k(z_c)} \right.$$
$$\left. + \sum_{k=1}^{\infty} \overline{O_k(z_0 - z_c) \widetilde{N}_k(z_c)} \right], \tag{5.47}$$

where the two sets of *moments* are:

$$\widetilde{M}_k(z_c) = \int_{S_c} I_{k-1}(z - z_c) n(z) u(z) \, dS(z), \quad \text{for } k \geq 1; \tag{5.48}$$

$$\widetilde{N}_1 = \int_{S_c} \left[n(z)\overline{u(z)} + \overline{n(z)} u(z) \right] dS(z); \tag{5.49}$$

$$\widetilde{N}_k(z_c) = \int_{S_c} \left\{ \overline{I_{k-1}(z - z_c)} \left[n(z)\overline{u(z)} + \overline{n(z)} u(z) \right] \right.$$
$$\left. - \overline{I_{k-2}(z - z_c)} \overline{z n(z) u(z)} \right\} dS(z), \quad \text{for } k \geq 2.$$

The M2M, M2L, and L2L translations remain the same for the T kernel integrals, except for the fact that $\widetilde{M}_0 = \widetilde{N}_0 = 0$. In fact, the moments M_k and $\widetilde{M}_k$ will be combined, as well as moments N_k and $\widetilde{N}_k$, so that only two sets of moments are involved in the M2M and M2L translations.

The local expansion for $D_u(z_0)$ is [cf. Eq. (4.68) for 2D elasticity]:

$$D_u(z_0) = \frac{1}{4\pi} \left[\sum_{l=0}^{\infty} L_l(z_L) I_l(z_0 - z_L) - z_0 \sum_{l=1}^{\infty} \overline{L_l(z_L) I_{l-1}(z_0 - z_L)} \right.$$
$$\left. + \sum_{l=0}^{\infty} K_l(z_L) \overline{I_l(z_0 - z_L)} \right], \tag{5.50}$$

where the coefficients $L_l(z_L)$ and $K_l(z_L)$ are given by Eqs. (5.43) and (5.44) with M_k replaced with $\widetilde{M}_k$ and N_k with $\widetilde{N}_k$.

5.4.6 Expansions for the Hypersingular Boundary Integral Equation

To derive the multipole expansions and local expansions for HBIE (5.25), we can simply apply the results for 2D elasticity problems and set Poisson's ratio to $\frac{1}{2}$ to obtain the results for 2D Stokes flow problems. The result of the local expansion for the first integral $F_t(z_0)$ in Eq. (5.35) for the HBIE is [cf. Eq. (4.69) for 2D elasticity]:

$$
F_t(z_0) = \frac{1}{4\pi} \left\{ \left[\sum_{l=0}^{\infty} L_{l+1}(z_L) I_l(z_0 - z_L) + \sum_{l=0}^{\infty} \overline{L_{l+1}(z_L) I_l(z_0 - z_L)} \right] n(z_0) \right.
$$
$$
\left. + \left[-z_0 \sum_{l=1}^{\infty} \overline{L_{l+1}(z_L) I_{l-1}(z_0 - z_L)} + \sum_{l=0}^{\infty} K_{l+1}(z_L) \overline{I_l(z_0 - z_L)} \right] \overline{n(z_0)} \right\},
$$
(5.51)

in which the expansion coefficients $L_l(z_L)$ and $K_l(z_L)$ are given by the same M2L translations in Eqs. (5.43) and (5.44), respectively. Therefore, the same sets of moments M_k and N_k used for $D_t(z_0)$ are used for $F_t(z_0)$ directly.

The local expansion for the second integral $F_u(z_0)$ in Eq. (5.36) for the HBIE is [cf. Eq. (4.70) for 2D elasticity]:

$$
F_u(z_0) = \frac{\mu}{2\pi} \left\{ \left[\sum_{l=0}^{\infty} L_{l+1}(z_L) I_l(z_0 - z_L) + \sum_{l=0}^{\infty} \overline{L_{l+1}(z_L) I_l(z_0 - z_L)} \right] n(z_0) \right.
$$
$$
\left. + \left[-z_0 \sum_{l=1}^{\infty} \overline{L_{l+1}(z_L) I_{l-1}(z_0 - z_L)} + \sum_{l=0}^{\infty} K_{l+1}(z_L) \overline{I_l(z_0 - z_L)} \right] \overline{n(z_0)} \right\},
$$
(5.52)

in which $L_l(z_L)$ and $K_l(z_L)$ are given by Eqs. (5.43) and (5.44) with M_k replaced with $\widetilde{M}_k$ and N_k with $\widetilde{N}_k$. Again, the same sets of moments $\widetilde{M}_k$ and $\widetilde{N}_k$ used for $D_u(z_0)$ are used for $F_u(z_0)$, and all the M2M, M2L, and L2L translations for the HBIE remain the same as those used for the CBIE.

The details of the fast multipole algorithms for solving 2D Stokes problems are similar to the ones for 2D potential and elasticity problems, which have been described in detail in the previous two chapters. Preconditioners for the fast multipole BEM are crucial for its convergence and computing efficiency. The block diagonal preconditioner is used in the study of the numerical examples, which are formed on each leaf by direct evaluations of the kernels on the elements within that leaf.

When constant elements (straight-line segment with one node) are used to discretize the BIEs, all of the moments can be evaluated analytically, as well as the integrations of the kernels in the near-field direct evaluations for 2D Stokes flow problems (Appendix A.3).

5.5 Fast Multipole Boundary Element Method for 3D Stokes Flow Problems

The fast multipole formulation for 3D Stokes flow problems can be derived readily from that for the 3D elasticity problems through a limiting process. For convenience and completeness, the results are subsequently summarized. First, we note that the fundamental solution in Eq. (5.18) for the 3D Stokes flow case can be written in the following form:

$$U_{ij}(\mathbf{x}, \mathbf{y}) = \frac{1}{8\pi\mu}\left(\delta_{ij}\frac{2}{r} - \frac{\partial}{\partial x_i}\frac{x_j - y_j}{r}\right). \tag{5.53}$$

Then, we start with the following expansion [same as Eq. (4.72)]:

$$\frac{1}{r(\mathbf{x}, \mathbf{y})} = \sum_{n=0}^{\infty}\sum_{m=-n}^{n}\overline{S_{n,m}}(\mathbf{x} - \mathbf{y}_c)R_{n,m}(\mathbf{y} - \mathbf{y}_c), \quad |\mathbf{y} - \mathbf{y}_c| < |\mathbf{x} - \mathbf{y}_c|, \tag{5.54}$$

where $\mathbf{y}_c$ is the expansion center close to the field point $\mathbf{y}$ and functions $R_{n,m}$ and $S_{n,m}$ are solid harmonic functions given in Eqs. (3.47) and (3.48), respectively. Substituting (5.54) into (5.53), we have:

$$U_{ij}(\mathbf{x}, \mathbf{y}) = \frac{1}{8\pi\mu}\sum_{n=0}^{\infty}\sum_{m=-n}^{n}\left[F_{ij,n,m}(\mathbf{x} - \mathbf{y}_c)\overline{R_{n,m}}(\mathbf{y} - \mathbf{y}_c)\right.$$
$$\left. + G_{i,n,m}(\mathbf{x} - \mathbf{y}_c)(\mathbf{y} - \mathbf{y}_c)_j\overline{R_{n,m}}(\mathbf{y} - \mathbf{y}_c)\right], \tag{5.55}$$

in which:

$$F_{ij,n,m}(\mathbf{x} - \mathbf{y}_c) = \delta_{ij}S_{n,m}(\mathbf{x} - \mathbf{y}_c) - (\mathbf{x} - \mathbf{y}_c)_j\frac{\partial}{\partial x_i}S_{n,m}(\mathbf{x} - \mathbf{y}_c), \tag{5.56}$$

$$G_{i,n,m}(\mathbf{x} - \mathbf{y}_c) = \frac{\partial}{\partial x_i}S_{n,m}(\mathbf{x} - \mathbf{y}_c). \tag{5.57}$$

Applying expression (5.55), with point $\mathbf{y}_c$ being close to subdomain S_c, we obtain the following *multipole expansion* for the U kernel integral in CBIE (5.21):

$$\int_{S_c} U_{ij}(\mathbf{x}, \mathbf{y})t_j(\mathbf{y})dS(\mathbf{y}) = \frac{1}{8\pi\mu}\sum_{n=0}^{\infty}\sum_{m=-n}^{n}\left[F_{ij,n,m}(\mathbf{x} - \mathbf{y}_c)\overline{M_{j,n,m}}(\mathbf{y}_c)\right.$$
$$\left. + G_{i,n,m}(\mathbf{x} - \mathbf{y}_c)\overline{M_{n,m}}(\mathbf{y}_c)\right], \tag{5.58}$$

in which:

$$M_{j,n,m}(\mathbf{y}_c) = \int_{S_c} R_{n,m}(\mathbf{y} - \mathbf{y}_c) t_j(\mathbf{y}) dS(\mathbf{y}),$$

$$M_{n,m}(\mathbf{y}_c) = \int_{S_c} (\mathbf{y} - \mathbf{y}_c)_j R_{n,m}(\mathbf{y} - \mathbf{y}_c) t_j(\mathbf{y}) dS(\mathbf{y}) \qquad (5.59)$$

are the *moments* for the given n and m. Note that these moments are similar to those for 3D elasticity problems.

To derive the multipole expansion for the T kernel integral in CBIE (5.21), we first note that the T kernel can be expressed as:

$$
\begin{aligned}
T_{ij} &= \Sigma_{ijk} n_k(\mathbf{y}) = [-P_i \delta_{jk} + \mu(U_{ij,k} + U_{ik,j})] n_k(\mathbf{y}) \\
&= [G_{,i} \delta_{jk} + \mu(U_{ij,k} + U_{ik,j})] n_k(\mathbf{y}) \\
&= \frac{1}{8\pi} \left[\delta_{ij} \frac{\partial}{\partial y_k} \left(\frac{1}{r} \right) + \delta_{ik} \frac{\partial}{\partial y_j} \left(\frac{1}{r} \right) + (\mathbf{y} - \mathbf{x})_j \frac{\partial^2}{\partial y_k \partial x_i} \left(\frac{1}{r} \right) \right. \\
&\quad \left. + (\mathbf{y} - \mathbf{x})_k \frac{\partial^2}{\partial y_j \partial x_i} \left(\frac{1}{r} \right) \right] n_k(\mathbf{y}), \qquad (5.60)
\end{aligned}
$$

where Eqs. (5.13) and (5.53) have been applied. Using the results in Eqs. (5.54) and (5.60), we obtain the multipole expansion for the T kernel integral in CBIE (5.21) as follows:

$$
\begin{aligned}
\int_{S_c} T_{ij}(\mathbf{x}, \mathbf{y}) u_j(\mathbf{y}) dS(\mathbf{y}) = \frac{1}{8\pi\mu} \sum_{n=0}^{\infty} \sum_{m=-n}^{n} & \left[F_{ij,n,m}(\mathbf{x} - \mathbf{y}_c) \overline{\tilde{M}_{j,n,m}}(\mathbf{y}_c) \right. \\
& \left. + G_{i,n,m}(\mathbf{x} - \mathbf{y}_c) \overline{\tilde{M}_{n,m}}(\mathbf{y}_c) \right], \qquad (5.61)
\end{aligned}
$$

in which:

$$\tilde{M}_{j,n,m}(\mathbf{y}_c) = \mu \int_{S_c} [n_k(\mathbf{y}) u_j(\mathbf{y}) + n_j(\mathbf{y}) u_k(\mathbf{y})] \frac{\partial}{\partial y_k} R_{n,m}(\mathbf{y} - \mathbf{y}_c) dS(\mathbf{y}),$$

$$\tilde{M}_{n,m}(\mathbf{y}_c) = \mu \int_{S_c} (\mathbf{y} - \mathbf{y}_c)_j [n_k(\mathbf{y}) u_j(\mathbf{y}) + n_j(\mathbf{y}) u_k(\mathbf{y})] \frac{\partial}{\partial y_k} R_{n,m}(\mathbf{y} - \mathbf{y}_c) dS(\mathbf{y}). \qquad (5.62)$$

We can also derive these moments from those for the 3D elasticity case by letting the Poisson's ratio tend to $\frac{1}{2}$ in the corresponding elasticity results. As in the elasticity case, only one in each of the pairs $(M_{j,n,m}, \tilde{M}_{j,n,m})$ and $(M_{n,m}, \tilde{M}_{n,m})$ is used in the moment calculations, depending on the BCs. Therefore, there are only four moments that need to be calculated on each boundary element.

When the expansion point is moved from $\mathbf{y}_c$ to $\mathbf{y}_{c'}$, we have the following M2M translations:

$$M_{j,n,m}(\mathbf{y}_{c'}) = \sum_{n'=0}^{n} \sum_{m'=-n'}^{n'} R_{n',m'}(\mathbf{y}_{c'} - \mathbf{y}_c) M_{j,n-n',m-m'}(\mathbf{y}_c),$$

$$M_{n,m}(\mathbf{y}_{c'}) = \sum_{n'=0}^{n} \sum_{m'=-n'}^{n'} R_{n',m'}(\mathbf{y}_{c'} - \mathbf{y}_c) \left[M_{n-n',m-m'}(\mathbf{y}_c) \right. \tag{5.63}$$
$$\left. + (\mathbf{y}_{c'} - \mathbf{y}_c)_j M_{j,n-n',m-m'}(\mathbf{y}_c) \right],$$

which also apply to $\widetilde{M}_{j,n,m}$ and $\widetilde{M}_{n,m}$.

The *local expansion* of the U kernel integral on S_c about the point $\mathbf{x} = \mathbf{x}_L$ is given by:

$$\int_{S_c} U_{ij}(\mathbf{x}, \mathbf{y}) t_j(\mathbf{y}) dS(\mathbf{y}) = \frac{1}{8\pi\mu} \sum_{n=0}^{\infty} \sum_{m=-n}^{n} \left[F_{ij,n,m}^R(\mathbf{x} - \mathbf{x}_L) L_{j,n,m}(\mathbf{x}_L) \right.$$
$$\left. + G_{i,n,m}^R(\mathbf{x} - \mathbf{x}_L) L_{n,m}(\mathbf{x}_L) \right], \tag{5.64}$$

where the local expansion coefficients are given by the following M2L translations:

$$L_{j,n,m}(\mathbf{x}_L) = (-1)^n \sum_{n'=0}^{\infty} \sum_{m'=-n'}^{n'} \overline{S_{n+n',m+m'}}(\mathbf{x}_L - \mathbf{y}_c) M_{j,n',m'}(\mathbf{y}_c),$$

$$L_{n,m}(\mathbf{x}_L) = (-1)^n \sum_{n'=0}^{\infty} \sum_{m'=-n'}^{n'} \overline{S_{n+n',m+m'}}(\mathbf{x}_L - \mathbf{y}_c) \left[M_{n',m'}(\mathbf{y}_c) \right. \tag{5.65}$$
$$\left. - (\mathbf{x}_L - \mathbf{y}_c)_j M_{j,n',m'}(\mathbf{y}_c) \right],$$

and $F_{ij,n,m}^R$ and $G_{i,n,m}^R$ are obtained from Eqs. (5.56) and (5.57), respectively, with $S_{n,m}$ replaced with $R_{n,m}$ in each case. The local expansion for the T kernel integral is similar to that of Eq. (5.64), only with $M_{j,n',m'}$ and $M_{n',m'}$ replaced with $\widetilde{M}_{j,n',m'}$ and $\widetilde{M}_{n',m'}$, respectively, in Eq. (5.65) when the local expansion coefficients are computed for the T kernel integral.

When the local expansion point is moved from $\mathbf{x}_L$ to $\mathbf{x}_{L'}$, we have the following L2L translations:

$$L_{j,n,m}(\mathbf{x}_{L'}) = \sum_{n'=n}^{\infty} \sum_{m'=-n'}^{n'} R_{n'-n,m'-m}(\mathbf{x}_{L'} - \mathbf{x}_L) L_{j,n',m'}(\mathbf{x}_L),$$

$$\tag{5.66}$$

$$L_{n,m}(\mathbf{x}_{L'}) = \sum_{n'=n}^{\infty} \sum_{m'=-n'}^{n'} R_{n'-n,m'-m}(\mathbf{x}_{L'} - \mathbf{x}_L) \left[L_{n',m'}(\mathbf{x}_L) - (\mathbf{x}_{L'} - \mathbf{x}_L)_j L_{j,n',m'}(\mathbf{x}_L) \right].$$

We can readily obtain the fast multipole formulation for the HBIE by taking the derivatives of the local expansions for the CBIE and invoking the constitutive equations.

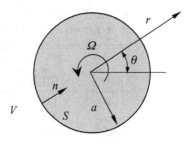

Figure 5.1. A rotating cylinder in an infinite fluid.

As in the 2D cases, we can readily obtain the fast multipole expansions for 3D Stokes flow problems by letting Poisson's ratio ν tend to $\frac{1}{2}$ in the corresponding results for 3D elasticity problems. All the M2M, M2L, and L2L translations are identical to those for the 3D elasticity case. Based on this fact, a fast multipole BEM program for 3D elasticity problems can be extended readily to a fast multipole BEM program for 3D Stokes flow problems if the direct BIE formulations are used for the Stokes flow problems.

5.6 Numerical Examples

Three examples are first presented in two dimensions to demonstrate the accuracy and the efficiency of the fast multipole BEM for Stokes flow problems. In all the 2D examples, the computations are done on a Pentium IV laptop PC with a 2.4-GHz CPU and 1-GB RAM. The numbers of terms in expansions are set to 20, the maximum number of elements in a leaf to 100, and the coupling constant $\beta = 1$ for the dual BIE. Finally, two 3D models are presented to further demonstrate the applications and potentials of the fast multipole BEM for solving Stokes flow problems.

5.6.1 Flow That Is Due to a Rotating Cylinder

The flow in an infinite 2D medium that is due to a rotating circular cylinder is considered first (Figure 5.1). The radius of the cylinder is a and the angular velocity is Ω. A solution to this problem exists [103]; that is, in the polar coordinate system, we have:

$$u_r(r, \theta) = 0, u_\theta(r, \theta) = \Omega a^2/r, \quad \sigma_{r\theta}(r, \theta) = -2\mu\Omega a^3/r^2, \quad (5.67)$$

which can be used to verify the BEM solutions. The velocity is specified on the boundary by use of the preceding results, and the tractions are sought with the BEM. For the fast multipole BEM solutions, the tolerance for convergence is set to 10^{-6}.

Table 5.1 shows the results of the tractions at the boundary computed by the fast multipole BEM with both the CBIE and CHBIE formulations (the HBIE cannot provide solutions in this case due to defect (3) mentioned in

Table 5.1. *Traction t_y at $(a, 0)$ and numbers of iterations used in the fast multipole BEM*

DOFs	t_y ($\times \mu\Omega a$)		Number of iterations	
	CBIE	CHBIE	CBIE	CHBIE
80	1.9999	1.9891	16	7
160	2.0003	1.9936	18	7
320	2.0054	1.9965	13	7
640	2.0028	1.9981	13	4
1280	2.0011	1.9990	14	4
2560	2.0005	1.9995	16	4
5120	1.9997	1.9998	21	4
10,240	1.9997	1.9999	28	4
20,480	2.0007	1.9999	32	4
Exact solution	2.0000			

Section 5.3). Although both BIE formulations give results of comparable accuracies, the CHBIE converges much faster than the CBIE, as indicated by the number of iterations used, which are also listed in Table 5.1.

Figure 5.2 is a plot of the traction components on the boundary of the cylinder with 40 elements and using CHBIE. Figure 5.3 shows the velocity computed at points inside the fluid domain with the same mesh and the

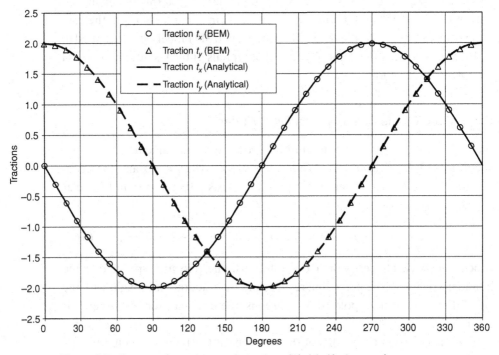

Figure 5.2. Computed tractions on boundary S (with 40 elements).

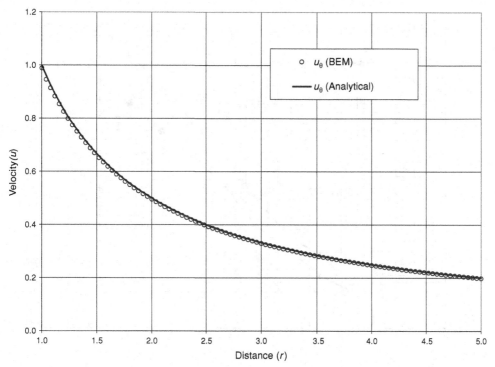

Figure 5.3. Computed velocity u_θ at points inside the fluid domain V (with 40 elements).

CHBIE. Both results demonstrate that the fast multipole BEM results are quite accurate with only 40 constant elements.

The CPU times used for the fast multipole BEM based on the CBIE and CHBIE approaches are plotted in Figure 5.4, which shows the significant advantage of the CHBIE formulation over the CBIE formulation. For example, for the model with 10,240 elements (DOFs = 20,480), the fast multipole BEM with CHBIE used about 17 s of CPU time, whereas the BEM with the CBIE used about 92 s, which is about four times slower. Higher condition numbers are observed for the CBIE and very low condition numbers for the CHBIE with a direct solver, which is consistent with the solution efficiency with the iterative solver.

5.6.2 Shear Flow Between Two Parallel Plates

The flow between two parallel plates (Figure 5.5) is studied next using the CBIE, HBIE, and CHBIE formulations. The top plate is moving with a constant speed v_0 in the x direction and a no-slip condition is assumed between the plates and fluid. The analytical solutions for this problem are:

$$u_x(x, y) = v_0 y/h, u_y = 0, \quad \sigma_x = \sigma_y = 0, \sigma_{xy} = \mu v_0/h. \qquad (5.68)$$

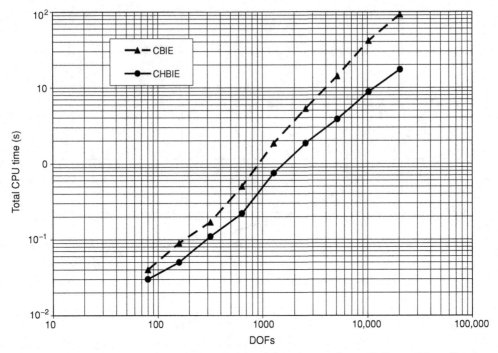

Figure 5.4. Total CPU time used for solving the rotating cylinder problem (log–log scale).

The purpose of this example is to show the behaviors of the BEM solutions as the ratio h/L approaches zero; that is, when the fluid domain becomes a narrow channel. The narrow spaces between two fingers of a MEMS comb-drive device closely resemble the configuration studied in this example with small ratios of h/L.

Mixed boundary conditions are used so that all of the three BIE formulations – that is, CBIE, HBIE, and CHBIE – can be tested. For the lower boundary, zero velocities are specified, whereas for the upper boundary, velocities

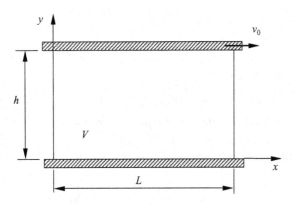

Figure 5.5. Shear flow between two parallel plates.

Table 5.2. *Comparison of the three BIE formulations for the shear flow problem*

h/L	Number of elements on edges L and h	Traction t_x ($\times \mu v_0/h$) at $L/2, 0$			Number of iterations		
		CBIE	HBIE	CHBIE	CBIE	HBIE	CHBIE
0	100/100	−0.99980	−1.00135	−0.99961	15	17	16
10^{-1}	100/20	−0.99998	−1.00264	−1.00185	25	21	21
10^{-2}	100/10	−1.00000	−1.00027	−1.00021	73	68	69
10^{-3}	100/5	−1.00000	−0.99985	−0.99988	142	67	67
10^{-4}	100/3	−1.00000	−0.99931	−0.99935	185	65	94
10^{-5}	100/2	−0.99998	−0.99943	−0.99514	227	49	70
10^{-6}	100/1	−0.99979	−0.99322	−0.98546	298	40	54
	Exact solutions		−1.00000				

are given as $u_x = v_0$ and $u_y = 0$. For the two vertical boundaries, tractions are given as $t_x = 0, t_y = \mu v_0/h$ at $x = L$ and $t_x = 0, t_y = -\mu v_0/h$ at $x = 0$. The tolerance for convergence in the fast multipole BEM is set to 10^{-6}.

Table 5.2 shows the dimension, discretization, and computed tractions at the midpoint of the lower boundary and the numbers of iterations used in the fast multipole BEM solutions with the three BIE formulations. It is observed that as the ratio of h/L becomes smaller, more iterations are needed for the CBIE formulation, whereas about the same numbers of iterations are needed for the HBIE and the CHBIE formulations. These results indicate the poor conditioning of the CBIE but also indicate the good and improved conditioning of the HBIE and CHBIE, respectively. Most interesting is the fact that even at $h/L = 10^{-6}$, all three BIE formulations can still provide reasonably good results of the tractions. The results from the HBIE and the CHBIE are slightly less accurate than those from the CBIE at small h/L, which may be caused by the extremely small elements on the two small vertical edges. Recall that for 2D problems, the finite part of the hypersingular integral is proportional to $1/R$, where R is the element length (see Appendix A.3). If R is very small, as is tested in this case, $1/R$ can be very large and can cause numerical errors in the BEM systems of equations. In fact, the BEM code fails when the ratio h/L is smaller than 10^{-6} for this example because of the existence of the hypersingular kernel H. This is different from the results reported in Chapter 3 and Ref. [57] for electrostatic MEMS problems, in which the ratio h/L of a beam can reach 10^{-16} for the dual BIE formulation that does not have the hypersingular kernel.

This example demonstrates that the dual BIE formulation can facilitate fast convergence for the fast multipole BEM even when the domain under consideration is extremely thin. This is consistent with the conclusions with the dual BIE approach for a fast multipole BEM in the context of electrostatic analysis of the MEMS models shown in Chapter 3 and Refs. [57, 59].

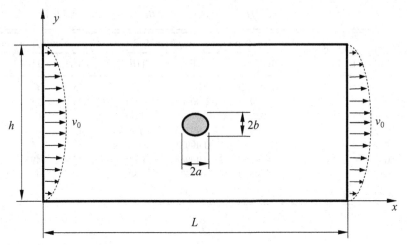

Figure 5.6. Channel flow around a cylinder.

5.6.3 Flow Through a Channel with Many Cylinders

We next consider an interior Dirichlet problem; that is, Stokes flows through a channel placed with one or multiple cylinders. The dimensions of the channel are shown in Figure 5.6. At the inlet of the channel ($x = 0$), the flow has a parabolic velocity profile:

$$u_x(0, y) = 4v_0(1 - y/h)y/h, \quad u_y(0, y) = 0, \qquad (5.69)$$

where v_0 is the maximum value of the velocity. At the outlet of the channel ($x = L$), the same velocity profile is assumed (see Figure 5.6); that is, the flow is assumed to have recovered from the disturbances by the cylinder(s) placed in the middle section of the channel. On the upper and lower boundaries and all cylinder boundaries, no-slip BCs are assumed. For this test, the tolerance for convergence for the solver is set to 10^{-5}.

First, the case with one circular cylinder placed in the center of the channel is studied, with $L = 2h$ and $a = 0.1h$. Figure 5.7 shows the velocity vector plot within the fluid obtained by use of the CHBIE. There are about 800 points distributed evenly inside the domain where the velocity is evaluated with the representation integral after the tractions are obtained from the BEM solutions.

Table 5.3 shows the total fluid force applied on the cylinder and evaluated by integration of the obtained traction field on the boundary of the cylinder (assuming a unit depth). There are 600 elements on the outer boundary and the number of elements on the cylinder increases. Both the CBIE and the CHBIE are used, and the results for the total force on the cylinder are very stable with the CBIE, whereas those with the CHBIE increase slowly to reach a stable value. The errors with the CHBIE may be due to the finite-part integrals in the HBIE on curved boundaries computed with constant elements that

Table 5.3. *Force F computed on the cylinder with CBIE and CHBIE*

Number of elements on cylinder	Total DOFs	Force F ($\times \mu v_0$)		No. of iterations		CPU time (s)	
		CBIE	CHBIE	CBIE	CHBIE	CBIE	CHBIE
320	1840	16.21	15.38	23	12	9.8	5.7
640	2480	16.21	15.76	26	12	16.2	8.5
1280	3760	16.21	15.96	28	9	30.8	12.0
2560	6320	16.21	16.06	28	9	68.8	26.2
5120	11,440	16.21	16.11	33	9	137.8	45.2
10,240	21,680	16.21	16.14	37	9	277.1	82.1

can introduce numerical errors. As shown in Table 5.3, the number of iterations with the CBIE increases as the model size increases, whereas the number of iterations with the CHBIE is almost constant and only about one half to one quarter of that for the CBIE.

Next, the models with multiple elliptic cylinders placed in the middle section of a channel with $L = 3h$ are studied. These models are motivated by the examples presented by Greengard *et al.* in Ref. [68], with different geometries, BCs, and numbers of elements.

Figure 5.8(a) shows the velocity plot for a 5×5 array of elliptic cylinders with a uniform distribution, and Figure 5.8(b) shows the velocity field with a random distribution, both using CHBIE with 16,600 DOFs. For the uniform distribution, 59 iterations are used (381-s CPU time), whereas for the random distribution, 82 iterations are used (491-s CPU time). It is observed that when more cylinders are placed in the same space or when cylinders are distributed randomly, the iteration numbers for the BEM solutions will increase because of the intensified interactions between the cylinders, as discussed in Ref. [40]. Figure 5.8(c) shows a larger model with 13×13 elliptic

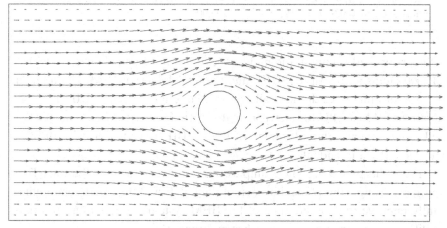

Figure 5.7. Vector plot of the velocity field for one circular cylinder with $a = 0.1h$ and $L = 2h$.

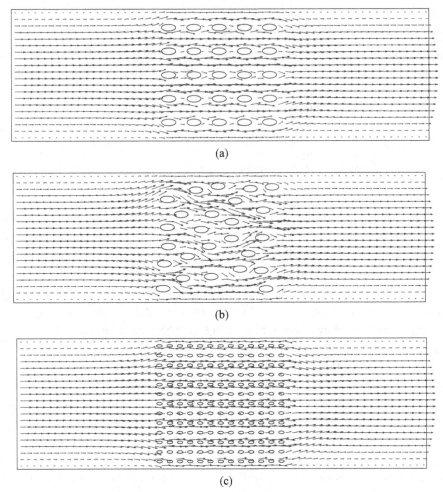

Figure 5.8. Various BEM models of the channel with many elliptic cylinders ($L = 3h$): (a) velocity plot for uniform distribution of 5×5 elliptic cylinders with $a = 0.05h$, $b = 0.5a$, DOFs $= 16{,}600$; (b) velocity plot for random distribution of 5×5 elliptic cylinders with $a = 0.05h$, $b = 0.5a$, DOFs $= 16{,}600$; (c) a larger model with 13×13 elliptic cylinders and $a = 0.02h$, $b = 0.5a$, DOFs $= 103{,}000$.

cylinders packed evenly in the middle section of the channel. The model has 103,000 DOFs and both the CBIE and the CHBIE are applied. The number of iterations increases dramatically for this large model. The CBIE used 248 iterations (9130-s CPU time), whereas the CHBIE used 168 iterations (6631-s CPU time). Again, the advantage of the CHBIE formulation with the fast multipole BEM is evident.

From the preceding 2D examples, we can conclude that the dual BIE approach, using a linear combination of the CBIE and the HBIE, can significantly improve the conditioning of the BEM systems of equations and thus facilitate faster convergence in the fast multipole BEM.

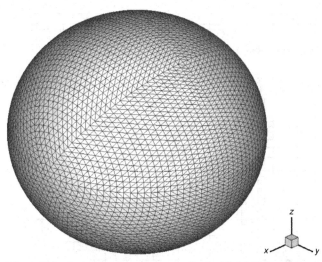

Figure 5.9. A translating sphere meshed with 10,800 constant triangular elements.

5.6.4 A Translating Sphere

We next study a 3D example by using a translating sphere, as shown in Figure 5.9. The sphere has radius R and moves with a constant velocity U_0 in the x direction in an infinite 3D fluid. The analytical solution of this Stokes flow problem is available and can be used to validate the BEM solutions.

Table 5.4 shows the computed total drag force on the sphere with different BEM discretizations and using the 3D fast multipole BEM code for Stokes

Table 5.4. *Computed drag force for the translating sphere*

		Drag force ($\times \mu R U_0$)		
Number of elements	Total DOFs	Stokes BIE	Elasticity BIE with $v = 0.499$	Number of iterations
432	1296	18.6585	18.6396	9
768	2304	18.7414	18.7226	11
1200	3600	18.7801	18.7613	12
2700	8100	18.8187	18.7999	12
4800	14,400	18.8319	18.8131	16
7500	22,500	18.8382	18.8194	17
10,800	32,400	18.8422	18.8234	16
14,700	44,100	18.8440	18.8252	18
19,200	57,600	18.8452	18.8264	20
24,300	72,900	18.8462	18.8274	20
30,000	90,000	18.8468	18.8280	20
36,300	108,900	18.8473	18.8285	20
43,200	129,600	18.8479	18.8291	20
Exact value		18.8496		

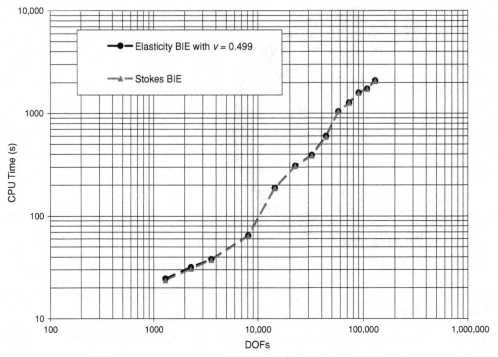

Figure 5.10. CPU times used for solving the translating sphere problem.

flow problems. For comparison, the results obtained with the 3D elasticity code with Poisson's ratio $\nu = 0.499$ (close to $\frac{1}{2}$) are also listed in the table. Both BIE results are very satisfactory compared with the analytical solution, with the Stokes BIE solutions closer to the analytical solution, as expected. The number of iterations is also listed, and fast convergence is observed for this exterior domain Stokes flow problem. In the BEM solutions, the tolerance for convergence is set to 10^{-4}, the number of terms in expansions to 15, and the number of elements in a leaf to 100. A Pentium D 3.2-GHz PC with 2-GB RAM is used for this study. The total CPU times used to run all the models are plotted in Figure 5.10, which shows a computational efficiency very close to $O(N)$.

5.6.5 Large-Scale Modeling of Multiple Particles

For the last example, a 3D model is shown with multipole particles that move through an infinite fluid, as shown in Figure 5.11. These particles are in the shape of a typical red-blood cell (RBC) and are used here to study the drag forces on these cells as they move in the fluid. The cells move with a constant velocity U_0 in the x direction. There are 40 cells in this model, and each cell is discretized with 7500 constant triangular elements (Figure 5.12). The number

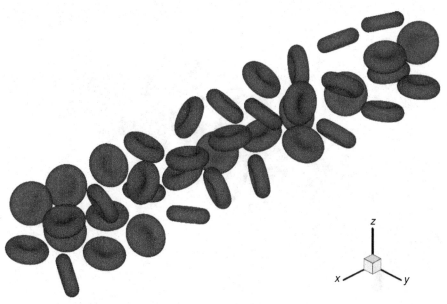

Figure 5.11. BEM model of 40 cells.

of the total DOFs for the model is 900,000, and the model is solved on the 2.4-GHz Pentium IV laptop with 1-GB RAM.

The computed drag forces (traction t_x) on the cells are shown in Figure 5.13. As expected, most of the values of traction are negative; that is, the forces are in the opposite direction of the motion of the cells. The large forces occur near the front and the end of the group of the cells. The CPU time used in solving this large BEM model with 900,000 DOFs is 730 min with 49 iterations for a tolerance of 10^{-4} and 15 expansion terms.

The results in this example are preliminary because the cells have been considered to be rigid bodies and the motion to be constant. These results can be regarded as a snapshot in one instance. To improve the BEM model, deformation of the cells should be considered; for example, by applying the

Figure 5.12. BEM mesh on each cell with 7500 constant triangular elements.

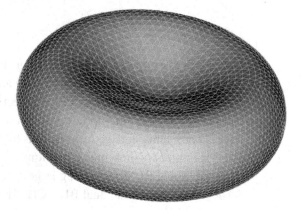

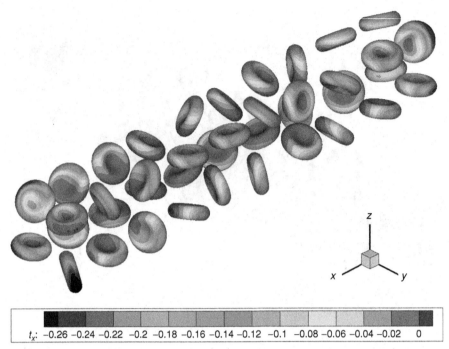

Figure 5.13. Computed traction (drag force) on the cells in the flow direction.

elasticity BEM or other mechanics models [103–105] for the cells. Quasi-dynamic analysis of the cells also can be conducted [40] to predict the evolution of the cell configurations.

Other potential applications of the 2D and the 3D Stokes flow fast multipole BEMs discussed in this chapter include studies of damping forces in MEMSs [95–98, 106] and Stokes flows interacting with deformable bodies [103–105]. Indeed, combining the Stokes flow fast multipole BEM code with the one for elasticity problems to study coupled fluid–structure interaction problems in general is an interesting research topic for applications in analyzing biological systems as well as MEMS devices.

5.7 Summary

In this chapter, the governing equations for solving Stokes flow problems are reviewed. The BIE formulations for 2D and 3D Stokes flow problems are presented, and the deficiencies of the direct CBIE and HBIE formulations are discussed. The fast multipole formulations for solving the BIEs are discussed for 2D problems, and the formulations for 3D problems are presented. Numerical examples are presented to demonstrate the accuracy and efficiencies of the fast multipole BEM for solving large-scale 2D and 3D Stokes flow problems. The advantages of the dual BIE (CHBIE) formulation for solving the Stokes flow problems are demonstrated regarding the computational efficiencies.

It is observed that the Stokes flow case is very similar to the elasticity case discussed in the previous chapter, regarding the fundamental solutions, singularities of the kernels, BIE formulations, fast multipole formulations, and solution procedures. Many of the results for Stokes flow problems can be obtained readily from their corresponding elasticity equations.

Problems

5.1. Solve the Stokes equation and obtain the analytical solutions in Eq. (5.67) for the rotating cylinder example.

5.2. Derive representation integral (5.22) for the pressure field $p(\mathbf{x})$.

5.3. Derive the K and H kernels in Eqs. (5.28) and (5.29), respectively, for 3D Stokes flow problems from the corresponding elasticity equations.

5.4. Verify expression (5.60) for the T kernel in three dimensions.

5.5. Verify multipole expansion (5.61) for the T kernel integral in the 3D CBIE.

5.6. Write a 2D Stokes flow conventional BEM code using the CBIE and constant elements, based on the code for 2D potential problems in Appendix B.1. Test your code on the 2D examples used in this chapter.

5.7. Write a 2D Stokes flow fast multipole BEM code using the CBIE and constant elements, based on the code for 2D potential problems in Appendix B.2. Test your code on the 2D examples used in this chapter and study its accuracy and efficiency.

6 Acoustic Wave Problems

Solving acoustic wave problems is one of the most important applications of the BEM, which can be used to predict sound fields for noise control in automobiles, airplanes, and many other consumer products. Acoustic waves often exist in an infinite medium outside a structure that is in vibration (a radiation problem) or impinged on by an incident wave (a scattering problem). With the BEM, only the boundary of the structure needs to be discretized. In addition, the BCs at infinity can be taken into account analytically in the BIE formulations, and thus these conditions are satisfied exactly. The governing equation for acoustic wave problems is the Helmholtz equation, which was solved using the BIE and BEM for more than four decades (see, e.g., some of the early work in Refs. [107–120]). Especially, the work by Burton and Miller in Ref. [108] is regarded as classical work that provides a very elegant way to overcome the so-called fictitious frequency difficulties existing in the conventional BIE for exterior acoustic wave problems. Burton and Miller's BIE formulation has been used by many others in their research on the BEM for acoustic problems (see, e.g., Refs. [50, 51, 121–125]).

The development of the fast multipole BEM for solving large-scale acoustic wave problems is perhaps the most important advance in the BEM that has made the BEM unmatched by other methods in modeling acoustic wave problems. The fast multipole method developed by Rokhlin and Greengard [33–35] has been extended to solving the Helmholtz equation for quite some time (see, e.g., Refs. [102, 126–137] and review [41]). Most of these works are good for solving acoustic wave problems at either low or high frequencies. For example, Greengard *et al.* [130] suggested a diagonal translation in the FMM for the low-frequency range. Rokhlin [127] proposed a diagonal form of the translation matrices for high-frequency ranges for the Helmholtz equation. Most recently, the same group also proposed integrated algorithms that are valid for a wide range of frequencies [137]. Gumerov and Duraiswami's research volume [136] is the first book that is devoted entirely to the topic of the FMM for solving Helmholtz equations in three dimensions.

In this chapter, we first review the basic governing equations for acoustic wave problems and the fundamental solutions. Then, the BIE formulations for acoustic wave problems are presented, followed by a discussion on the fast multipole BEM for solving the acoustic BIEs in both two and three dimensions. Finally, several numerical examples are presented to demonstrate the advantages of the fast multipole BEM in solving acoustic wave problems and to discuss the remaining challenges.

6.1 Basic Equations in Acoustics

In 1D space, the acoustic wave equation can be written as:

$$\frac{\partial^2 \phi}{\partial x^2} - \frac{1}{c^2} \frac{\partial^2 \phi}{\partial t^2} + Q\delta(x, x_Q) = 0, \tag{6.1}$$

in which $\phi = \phi(x, t)$ is the perturbation acoustic pressure, x is the coordinate, t is the time, c is the speed of sound in the medium (e.g., $c = 343.6$ m/s in air at a temperature of 20°C), and $Q\delta(x, x_Q)$ represents a possible point source located at x_Q. The preceding equation can be applied to describe the acoustic wave in a 1D long duct that is due to a disturbance along the axis of the duct (x direction).

We can verify that the solutions of Eq. (6.1) are of the following forms when $Q = 0$:

$$\phi(x, t) = f(x - ct), \quad \phi(x, t) = g(x + ct). \tag{6.2}$$

Function $f(x - ct)$ represents a right-traveling wave (waves moving along the $+x$ direction) and $g(x + ct)$ is a left-traveling wave (waves moving along the $-x$ direction).

Similarly, in 2D or 3D spaces, the acoustic wave equation can be written as:

$$\nabla^2 \phi - \frac{1}{c^2} \frac{\partial^2 \phi}{\partial t^2} + Q\delta(\mathbf{x}, \mathbf{x}_Q) = 0, \quad \forall \mathbf{x} \in E, \tag{6.3}$$

where $\phi = \phi(\mathbf{x}, t)$ is the perturbation acoustic pressure at point $\mathbf{x}$ and time t, c is the speed of sound, $Q\delta(\mathbf{x}, \mathbf{x}_Q)$ is a typical point source located at $\mathbf{x}_Q$ in E (Figure 6.1), and $\nabla^2( ) = \partial^2( )/\partial x_k \partial x_k = ( )_{,kk}$ is the Laplace operator. The acoustic domain E is considered to be isotropic and homogeneous and can be an *infinite* domain exterior to a body V or a *finite* domain interior to a closed surface.

For time harmonic waves, the point source intensity $Q = \tilde{Q}e^{-i\omega t}$ and the solutions to the governing equation can be written as:

$$\phi(\mathbf{x}, t) = \tilde{\phi}(\mathbf{x}, \omega)e^{-i\omega t}, \tag{6.4}$$

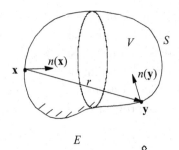

Figure 6.1. The acoustic medium E, body V, and boundary S.

in which $\tilde{\phi}(\mathbf{x}, \omega)$ is the (complex) acoustic pressure in the frequency domain, ω is the circular frequency, and $i = \sqrt{-1}$. Substituting Eq. (6.4) into Eq. (6.3), we obtain:

$$\nabla^2 \tilde{\phi} + \frac{\omega^2}{c^2} \tilde{\phi} + \tilde{Q}\delta(\mathbf{x}, \mathbf{x}_Q) = 0, \quad \forall \mathbf{x} \in E.$$

Let $k = \omega/c$ be the *wavenumber* and, for convenience, we drop the tildes in the preceding equation. We obtain the following governing equation for acoustic wave problems:

$$\nabla^2 \phi + k^2 \phi + Q\delta(\mathbf{x}, \mathbf{x}_Q) = 0, \quad \forall \mathbf{x} \in E, \tag{6.5}$$

with $\phi(\mathbf{x}, \omega)$ being the complex acoustic pressure and $Q\delta(\mathbf{x}, \mathbf{x}_Q)$ representing the point source located at $\mathbf{x}_Q$ (inside domain E). The preceding equation is the well-known inhomogeneous *Helmholtz equation*. Note that the cyclic frequency:

$$f = \omega/2\pi \quad \text{(with units of inverse seconds, or Hertz)}, \tag{6.6}$$

and the wavelength:

$$\lambda = c/f = 2\pi c/\omega = 2\pi/k \quad \text{(with units of meters)}. \tag{6.7}$$

The BCs for the governing equation can be classified as follows:

Pressure is given:

$$\phi = \bar{\phi}, \quad \forall \mathbf{x} \in S; \tag{6.8}$$

Velocity is given:

$$q \equiv \frac{\partial \phi}{\partial n} = \bar{q} = i\omega\rho v_n, \quad \forall \mathbf{x} \in S; \tag{6.9}$$

Impedance is given:

$$\phi = Z v_n, \quad \forall \mathbf{x} \in S, \tag{6.10}$$

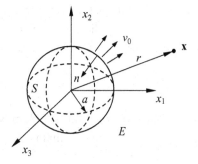

Figure 6.2. A pulsating sphere in an infinite acoustic domain E.

where ρ is the mass density (e.g., $\rho = 1.29$ kg/m^3 for air at 0°C and 1-atm pressure), v_n is the normal velocity, Z is the specific acoustic impedance, and the quantities with overbars indicate given values.

For the boundary-value problem for acoustic waves, we need to solve governing equation (6.5) at a given frequency or wavenumber and under the BCs in (6.8)–(6.10). Once we have the solution $\phi(\mathbf{x},\omega)$ in the frequency domain, we can obtain the solution in the time domain from Eq. (6.4) for the time harmonic case.

There are two typical types of problems in acoustic wave analysis. One is called a *radiation problem*, in which a structure is in vibration and causes disturbances in the acoustic field outside or inside the structure. In this case, the velocity on the boundary S is specified in the acoustic analysis. Another type of acoustic wave problem is called a *scattering problem*, in which the structure stands still and an incoming disturbance (a plane incident wave or an incident wave from a point source) interacts with the structure and waves are scattered by the structure.

For exterior (infinite domain) acoustic wave problems, in addition to the boundary conditions on S, the field at infinity must satisfy the following *Sommerfeld radiation condition*:

$$\lim_{R \to \infty} \left[R \left| \frac{\partial \phi}{\partial R} - ik\phi \right| \right] = 0, \tag{6.11}$$

where R is the radius of a large sphere covering the structure and ϕ is the radiated wave in a radiation problem or the scattered wave in a scattering problem. Basically, the Sommerfeld condition says that any acoustic disturbances caused by the structure (either radiated or scattered) should die out at infinity based on the energy considerations.

As an example, let us consider *a pulsating sphere* in an infinite acoustic medium (Figure 6.2). This example is often used as a test problem to verify the numerical solutions. The sphere has a radius a and is applied with a uniform velocity v_0 in the radial direction (imagine a balloon expanding and contracting uniformly in the radial directions and harmonically in time). This is a radiation problem, and the boundary-value problem is given by:

$$\nabla^2\phi + k^2\phi = 0 \quad \forall \mathbf{x} \in E,$$

$$\frac{\partial\phi}{\partial n} = -i\omega\rho v_0, \quad \forall \mathbf{x} \in S. \tag{6.12}$$

To solve this problem, we note the spherical symmetry and use the spherical coordinates; that is, $\phi(\mathbf{x}, \omega) = \phi(r, \omega)$, where r is the radial coordinate (see Figure 6.2).

We have in the spherical coordinate system:

$$\nabla^2\phi = \left(\frac{\partial^2}{\partial r^2} + \frac{2}{r}\frac{\partial}{\partial r}\right)\phi = \frac{1}{r}\frac{d^2}{dr^2}(r\phi);$$

therefore, the governing equation in (6.12) is reduced to:

$$\frac{d^2}{dr^2}(r\phi) + k(r\phi) = 0. \tag{6.13}$$

The solution of this ODE is of the form:

$$\phi(r, \omega) = \frac{1}{r}\left(A_1 e^{ikr} + A_2 e^{-ikr}\right),$$

where A_1 and A_2 are two constants. For the field caused by the pulsating sphere, the wave should be an outgoing wave (traveling away from the sphere). Thus, the second term representing an incoming wave (traveling toward the sphere) should vanish ($A_2 = 0$). We have:

$$\phi(r, \omega) = \frac{1}{r}A_1 e^{ikr}.$$

To determine A_1, we apply the BC in Eq. (6.12) and find that:

$$A_1 = \frac{i\omega\rho v_0 a^2}{ika - 1}e^{-ika} = \frac{\rho c v_0(ika)}{ika - 1}ae^{-ika}.$$

Therefore, the solution to the pulsating-sphere problem is found to be:

$$\phi(r, \omega) = \frac{\rho c v_0(ika)}{ika - 1}\frac{a}{r}e^{ik(r-a)}, \tag{6.14}$$

which represents the perturbation pressure field that is due to the pulsating sphere. We notice that this is an outgoing wave and it vanishes at infinity (i.e., satisfies the Sommerfeld radiation condition).

6.2 Fundamental Solution for Acoustic Wave Problems

If we place a unit concentrated source (disturbance or "pulsating point") at point $\mathbf{x}$ in an acoustic medium occupying the full space, then the mathematical representation for the response (acoustic disturbance pressure) at another point $\mathbf{y}$ is called the fundamental solution or the full-space Green's function for acoustic problems. This fundamental solution, denoted as $G(\mathbf{x}, \mathbf{y}, \omega)$ in this chapter, satisfies the following governing equation:

$$\nabla^2 G(\mathbf{x}, \mathbf{y}, \omega) + k^2 G(\mathbf{x}, \mathbf{y}, \omega) + \delta(\mathbf{x}, \mathbf{y}) = 0, \quad \forall \mathbf{x}, \mathbf{y} \in R^2/R^3, \tag{6.15}$$

in which the derivative is taken at field point $\mathbf{y}$ and the Dirac δ function represents the unit source at source point $\mathbf{x}$. $G(\mathbf{x}, \mathbf{y}, \omega)$ should represent an outgoing wave and have spherical (radial) symmetry. From the solution in Eq. (6.14) for the pulsating sphere, we know that $G(\mathbf{x}, \mathbf{y}, \omega)$ should have the following form for 3D problems:

$$G(\mathbf{x}, \mathbf{y}, \omega) = \frac{A}{r} e^{ikr}, \tag{6.16}$$

where r is the distance between $\mathbf{x}$ and $\mathbf{y}$ and A is a constant. To determine A, which represents the strength of the source, we integrate Eq. (6.15) over a small spherical domain $E_\varepsilon(\mathbf{x})$ centered at $\mathbf{x}$, with radius ε and boundary $S_\varepsilon(\mathbf{x})$ to obtain:

$$\int_{E_\varepsilon(\mathbf{x})} \left[\nabla^2 G + k^2 G\right] dE(\mathbf{y}) = -\int_{E_\varepsilon(\mathbf{x})} \delta(\mathbf{x}, \mathbf{y}) dE(\mathbf{y}) = -1. \tag{6.17}$$

Applying the Gauss theorem and expression (6.16), we have:

$$\int_{E_\varepsilon(\mathbf{x})} \nabla^2 G \, dE(\mathbf{y}) = \int_{S_\varepsilon(\mathbf{x})} \frac{\partial G}{\partial n} dS(\mathbf{y}) = 4\pi A(ik\varepsilon - 1)e^{ik\varepsilon}.$$

Similarly, using the spherical coordinates (r, θ, φ), we obtain:

$$\int_{E_\varepsilon(\mathbf{x})} k^2 G \, dE(\mathbf{y}) = k^2 A \int_0^\varepsilon \int_0^{2\pi} \int_0^\pi \frac{1}{r} e^{ikr} r^2 \sin\varphi \, d\varphi \, d\theta \, dr$$

$$= 4\pi A \left[(1 - ik\varepsilon)e^{ik\varepsilon} - 1\right].$$

Substituting the preceding two results into Eq. (6.17), we obtain $A = 1/4\pi$ and the following results for the fundamental solution for 3D acoustic wave problems:

$$G(\mathbf{x}, \mathbf{y}, \omega) = \frac{1}{4\pi r} e^{ikr}, \tag{6.18}$$

$$F(\mathbf{x}, \mathbf{y}, \omega) \equiv \frac{\partial G(\mathbf{x}, \mathbf{y}, \omega)}{\partial n(\mathbf{y})} = \frac{1}{4\pi r^2}(ikr - 1)r_{,j} \, n_j(\mathbf{y})e^{ikr}. \tag{6.19}$$

Note that when $k = 0$ – that is, when the problem becomes a static one – the preceding two results are exactly the same as those for the 3D potential problems discussed in Chapter 2. This is expected because Helmholtz equation (6.5) becomes a Poisson equation if $k = 0$.

For 2D acoustic wave problems, we have the following results for the fundamental solution:

$$G(\mathbf{x}, \mathbf{y}, \omega) = \frac{i}{4} H_0^{(1)}(kr), \tag{6.20}$$

$$F(\mathbf{x}, \mathbf{y}, \omega) \equiv \frac{\partial G(\mathbf{x}, \mathbf{y}, \omega)}{\partial n(\mathbf{y})} = -\frac{ik}{4} H_1^{(1)}(kr) r_{,l}\, n_l(\mathbf{y}), \tag{6.21}$$

in which $H_n^{(1)}(\ )$ denotes the Hankel function of the first kind [138].

6.3 Boundary Integral Equation Formulations

To derive the BIE corresponding to Helmholtz equation (6.5), we apply the second Green's identity given in Eq. (1.24) (we use the 3D case as the example):

$$\int_E \left[u\nabla^2 v - v\nabla^2 u \right] dE = \int_{S \cup S_R} \left[u \frac{\partial v}{\partial n} - v \frac{\partial u}{\partial n} \right] dS, \tag{6.22}$$

in which E is a domain bounded by the boundary S of the structure (see Figure 6.1) and a large sphere S_R of radius R (with $R \to \infty$). For interior problems, S_R does not exist. Let $v(\mathbf{y}) = \phi(\mathbf{y})$, which satisfies Eq. (6.5), and $u(\mathbf{y}) = G(\mathbf{x}, \mathbf{y}, \omega)$, which satisfies Eq. (6.15). We have from identity (6.22):

$$\int_E \left[G\nabla^2 \phi - \phi\nabla^2 G \right] dE = \int_{S \cup S_R} \left[G \frac{\partial \phi}{\partial n} - \phi \frac{\partial G}{\partial n} \right] dS.$$

Applying Eqs. (6.5), (6.15), and (1.25), we obtain:

$$\phi(\mathbf{x}) = \int_{S \cup S_R} \left[G(\mathbf{x}, \mathbf{y}, \omega) q(\mathbf{y}) - F(\mathbf{x}, \mathbf{y}, \omega)\phi(\mathbf{y}) \right] dS(\mathbf{y}) + QG(\mathbf{x}, \mathbf{x}_Q, \omega), \quad \forall \mathbf{x} \in E, \tag{6.23}$$

in which $q = \partial\phi/\partial n$ and the term $QG(\mathbf{x}, \mathbf{x}_Q, \omega)$ is due to the point source at $\mathbf{x}_Q$ in the domain.

Now, consider the integral on S_R as $R \to \infty$ for an infinite domain. For this purpose, we first note the following inequalities:

$$\left| \int f(x) dx \right| \le \int |f(x)|\, dx,$$

$$|f(x) + g(x)| \le |f(x)| + |g(x)|, \text{ and so on.}$$

For radiation problems, ϕ is the radiated wave, and we evaluate:

$$\lim_{R \to \infty} \left| \int_{S_R} \left[G(\mathbf{x}, \mathbf{y}, \omega) q(\mathbf{y}) - F(\mathbf{x}, \mathbf{y}, \omega)\phi(\mathbf{y}) \right] dS(\mathbf{y}) \right|$$

$$\le \lim_{R \to \infty} \int_{S_R} |Gq - F\phi|\, dS \le \lim_{R \to \infty} \int_{S_R} \left| \left[\frac{1}{4\pi R} \frac{\partial \phi}{\partial R} - \frac{(ikR - 1)}{4\pi R^2}\phi \right] e^{ikR} \right| dS$$

$$\le \lim_{R \to \infty} \int_{S_R} \left[\frac{1}{4\pi R} \left| \frac{\partial \phi}{\partial R} - ik\phi \right| + \frac{1}{4\pi R^2} |\phi| \right] dS$$

$$= \lim_{R \to \infty} R \left| \frac{\partial \phi}{\partial R} - ik\phi \right| + \lim_{R \to \infty} |\phi| = 0 \tag{6.24}$$

by using the Sommerfeld radiation condition in Eq. (6.11) and noting the fact that ϕ itself should vanish at the infinity.

For scattering problems, ϕ is the total wave that is the sum of the incident wave ϕ^I and the scattered wave ϕ^S; that is, $\phi = \phi^I + \phi^S$. The scattered wave ϕ^S also satisfies the Sommerfeld condition. Thus, for scattering problems, we have for the integral on S_R as $R \to \infty$:

$$\int_{S_R} [G(\mathbf{x}, \mathbf{y}, \omega)q(\mathbf{y}) - F(\mathbf{x}, \mathbf{y}, \omega)\phi(\mathbf{y})] \, dS(\mathbf{y})$$

$$= \int_{S_R} [Gq^I - F\phi^I] \, dS + \int_{S_R} [Gq^S - F\phi^S] \, dS,$$

in which $q^I = \partial\phi^I/\partial n$ and $q^S = \partial\phi^S/\partial n$. The first integral on the right-hand side is equal to the incident wave ϕ^I by considering ϕ^I within the domain as enclosed by S_R, and the second integral vanishes as in the radiation problems. Therefore, for the scattering problems, we have:

$$\int_{S_R} [G(\mathbf{x}, \mathbf{y}, \omega)q(\mathbf{y}) - F(\mathbf{x}, \mathbf{y}, \omega)\phi(\mathbf{y})] \, dS(\mathbf{y}) = \phi^I(\mathbf{x}). \qquad (6.25)$$

From the results in Eqs. (6.24) and (6.25), we obtain from Eq. (6.23) the following general representation integral:

$$\phi(\mathbf{x}) = \int_S [G(\mathbf{x}, \mathbf{y}, \omega)q(\mathbf{y}) - F(\mathbf{x}, \mathbf{y}, \omega)\phi(\mathbf{y})] \, dS(\mathbf{y})$$

$$+ \phi^I(\mathbf{x}) + QG(\mathbf{x}, \mathbf{x}_Q, \omega), \quad \forall \mathbf{x} \in E, \qquad (6.26)$$

where the incident wave $\phi^I(\mathbf{x})$ does not present for radiation problems.

Equation (6.26) is the representation integral of the solution ϕ inside domain E for Helmholtz equation (6.5) for both exterior and interior domain problems. Once the values of both ϕ and q are known on S, Eq. (6.26) can be applied to calculate ϕ everywhere in E, if needed.

Let the source point $\mathbf{x}$ approach the boundary S. We obtain the following CBIE for acoustic wave problems:

$$c(\mathbf{x})\phi(\mathbf{x}) = \int_S [G(\mathbf{x}, \mathbf{y}, \omega)q(\mathbf{y}) - F(\mathbf{x}, \mathbf{y}, \omega)\phi(\mathbf{y})] \, dS(\mathbf{y})$$

$$+ \phi^I(\mathbf{x}) + QG(\mathbf{x}, \mathbf{x}_Q, \omega), \quad \forall \mathbf{x} \in S, \qquad (6.27)$$

where the constant $c(\mathbf{x}) = \frac{1}{2}$ if S is smooth around $\mathbf{x}$. This CBIE can be used to solve for the unknown ϕ and q on S. The integral with the G kernel is a weakly singular integral, whereas the one with the F kernel is a strongly singular (CPV) integral, as in the potential case.

It is well known that this CBIE has a major defect for exterior domain problems; that is, it has nonunique solutions at a set of fictitious eigenfrequencies associated with the resonate frequencies of the corresponding interior problems [108]. This difficulty is referred to as the *fictitious eigenfrequency difficulty*. A remedy to this problem is to use the normal derivative BIE in conjunction with this CBIE. Taking the derivative of integral representation (6.26) with respect to the normal at the point $\mathbf{x}$ and letting $\mathbf{x}$ approach S, we obtain the following HBIE for acoustic wave problems:

$$\tilde{c}(\mathbf{x})q(\mathbf{x}) = \int_S [K(\mathbf{x}, \mathbf{y}, \omega)q(\mathbf{y}) - H(\mathbf{x}, \mathbf{y}, \omega)\phi(\mathbf{y})]\, dS(\mathbf{y})$$

$$+ q^I(\mathbf{x}) + QK(\mathbf{x}, \mathbf{x}_Q, \omega), \quad \forall \mathbf{x} \in S, \tag{6.28}$$

where $\tilde{c}(\mathbf{x}) = \frac{1}{2}$ if S is smooth around $\mathbf{x}$. For 3D problems, the two new kernels are given by:

$$K(\mathbf{x}, \mathbf{y}, \omega) \equiv \frac{\partial G(\mathbf{x}, \mathbf{y}, \omega)}{\partial n(\mathbf{x})} = -\frac{1}{4\pi r^2}(ikr - 1)r_{,j}\, n_j(\mathbf{x})e^{ikr}, \tag{6.29}$$

$$H(\mathbf{x}, \mathbf{y}, \omega) \equiv \frac{\partial F(\mathbf{x}, \mathbf{y}, \omega)}{\partial n(\mathbf{x})} = \frac{1}{4\pi r^3}\left\{(1 - ikr)n_j(\mathbf{y})\right.$$

$$\left. + \left[k^2r^2 - 3(1 - ikr)\right]r_{,j}\, r_{,l}\, n_l(\mathbf{y})\right\} n_j(\mathbf{x})e^{ikr}, \tag{6.30}$$

and for 2D problems, the two new kernels are:

$$K(\mathbf{x}, \mathbf{y}, \omega) \equiv \frac{\partial G(\mathbf{x}, \mathbf{y}, \omega)}{\partial n(\mathbf{x})} = \frac{ik}{4}H_1^{(1)}(kr)r_{,j}\, n_j(\mathbf{x}), \tag{6.31}$$

$$H(\mathbf{x}, \mathbf{y}, \omega) \equiv \frac{\partial F(\mathbf{x}, \mathbf{y}, \omega)}{\partial n(\mathbf{x})} = \frac{ik}{4r}H_1^{(1)}(kr)n_j(\mathbf{x})n_j(\mathbf{y})$$

$$- \frac{ik^2}{4}H_2^{(1)}(kr)r_{,j}\, n_j(\mathbf{x})r_{,l}\, n_l(\mathbf{y}). \tag{6.32}$$

In HBIE (6.28), the integral with the K kernel is a strongly singular (CPV) integral, whereas the one with the H kernel is a hypersingular (HFP) integral.

For exterior acoustic wave problems, a dual BIE (CHBIE or composite BIE [50]) formulation using a linear combination of CBIE (6.27) and HBIE (6.28) can be written as:

$$\text{CBIE} + \beta\, \text{HBIE} = 0, \tag{6.33}$$

where β is the coupling constant. This dual BIE formulation is called the Burton–Miller formulation [108] for acoustic wave problems and was shown by Burton and Miller to yield unique solutions at all frequencies, if β is a complex number (which, for example, can be chosen as $\beta = i/k$ [112, 114, 122, 139]).

6.4 Weakly Singular Forms of the Boundary Integral Equations

CBIE (6.27) and HBIE (6.28) contain singular integrals that are difficult to evaluate analytically on higher-order elements. Numerical integration can be used to compute all the singular integrals with proper care, but it was found to be not very efficient computationally with higher-order elements. As in all of the other problems using the BIE and BEM, the best approach in such cases is to use the weakly singular forms of these BIEs, which are obtained analytically and do not introduce any approximations. For dynamic problems, however, there were no integral identities found with the dynamic kernels directly. Therefore, there are extra steps in the development of the weakly singular forms of the BIEs for acoustic wave and other dynamic problems using the BIEs [50, 52, 124].

For CBIE (6.27), we first note that the free-term coefficient $c(\mathbf{x})$ can be written as (see notation used in Chapter 2):

$$c(\mathbf{x}) = 1 + \lim_{\varepsilon \to 0} \int_{S_\varepsilon(\mathbf{x})} F(\mathbf{x}, \mathbf{y}, \omega) dS(\mathbf{y}) = 1 + \lim_{\varepsilon \to 0} \int_{S_\varepsilon(\mathbf{x})} \overline{F}(\mathbf{x}, \mathbf{y}) dS(\mathbf{y})$$

$$= \gamma - \lim_{\varepsilon \to 0} \int_{S-S_\varepsilon(\mathbf{x})} \overline{F}(\mathbf{x}, \mathbf{y}) dS(\mathbf{y})$$

$$= \gamma - \int_S \overline{F}(\mathbf{x}, \mathbf{y}) dS(\mathbf{y}), \quad \forall \mathbf{x} \in S \quad \text{(a CPV integral)}, \tag{6.34}$$

in which $\overline{F}(\mathbf{x}, \mathbf{y}) = F(\mathbf{x}, \mathbf{y}, 0)$ is the static F kernel for potential problems (an overbar is added in this chapter to distinguish static kernels from the dynamic ones), $\gamma = 0$ for a finite domain, and $\gamma = 1$ for an infinite domain. In deriving Eq. (6.34), the first identity in Eq. (2.7) for the potential (static) kernel is applied. The fact that the dynamic kernel can be replaced with the static kernel is due to the following results for small r (with the 3D case as the example):

$$G(\mathbf{x}, \mathbf{y}, \omega) - \overline{G}(\mathbf{x}, \mathbf{y}) = \frac{1}{4\pi r} \left[e^{ikr} - 1 \right]$$

$$= \frac{1}{4\pi r} \left[ikr + \frac{1}{2!}(ikr)^2 + \frac{1}{3!}(ikr)^3 + \cdots + \right]$$

$$= a_0 + a_1 r + a_2 r^2 + \cdots +,$$

$$F(\mathbf{x}, \mathbf{y}, \omega) - \overline{F}(\mathbf{x}, \mathbf{y}) = \frac{\partial}{\partial n(\mathbf{y})} \left[a_0 + a_1 r + a_2 r^2 + \cdots + \right]$$

$$= O(r_{,n}) = O(r), \quad \text{as } r \to 0, \tag{6.35}$$

where $a_0, a_1, a_2, \ldots$, are some constants. Substituting the expression in (6.34) for $c(\mathbf{x})$ into CBIE (6.27), we obtain the following weakly singular form of the CBIE for acoustic wave problems:

$$\gamma \phi(\mathbf{x}) + \int_S \left[F(\mathbf{x}, \mathbf{y}, \omega) - \overline{F}(\mathbf{x}, \mathbf{y}) \right] \phi(\mathbf{y}) dS(\mathbf{y}) + \int_S \overline{F}(\mathbf{x}, \mathbf{y}) \left[\phi(\mathbf{y}) - \phi(\mathbf{x}) \right] dS(\mathbf{y})$$

$$= \int_S G(\mathbf{x}, \mathbf{y}, \omega) q(\mathbf{y}) dS(\mathbf{y}) + \phi^I(\mathbf{x}) + QG(\mathbf{x}, \mathbf{x}_Q, \omega), \quad \forall \mathbf{x} \in S, \tag{6.36}$$

in which all three integrals are now, at most, weakly singular and can be handled readily by numerical integration schemes.

Similarly, if we introduce the static kernel and a two-term subtraction and apply the identities satisfied by the static kernels, we can show that HBIE (6.28) can be written in the following weakly singular form [50, 51]:

$$
\gamma q(\mathbf{x}) + \int_S \left[H(\mathbf{x}, \mathbf{y}, \omega) - \overline{H}(\mathbf{x}, \mathbf{y}) \right] \phi(\mathbf{y}) dS(\mathbf{y})
$$

$$
+ \int_S \overline{H}(\mathbf{x}, \mathbf{y}) \left[\phi(\mathbf{y}) - \phi(\mathbf{x}) - \frac{\partial \phi}{\partial \xi_\alpha}(\mathbf{x})(\xi_\alpha - \xi_{o\alpha}) \right] dS(\mathbf{y})
$$

$$
+ e_{\alpha k} \frac{\partial \phi}{\partial \xi_\alpha}(\mathbf{x}) \int_S \left[\overline{K}(\mathbf{x}, \mathbf{y}) n_k(\mathbf{y}) + \overline{F}(\mathbf{x}, \mathbf{y}) n_k(\mathbf{x}) \right] dS(\mathbf{y})
$$

$$
= \int_S \left[K(\mathbf{x}, \mathbf{y}, \omega) + \overline{F}(\mathbf{x}, \mathbf{y}) \right] q(\mathbf{y}) dS(\mathbf{y})
$$

$$
- \int_S \overline{F}(\mathbf{x}, \mathbf{y}) [q(\mathbf{y}) - q(\mathbf{x})] dS(\mathbf{y}) + q^I(\mathbf{x}) + QK(\mathbf{x}, \mathbf{x}_Q, \omega), \quad \forall \mathbf{x} \in S,
$$

$$(6.37)$$

in which ξ_α ($\alpha = 1$ for two dimensions and $\alpha = 1, 2$ for three dimensions) are local coordinates in tangential directions at $\mathbf{x} \in S$ and $e_{\alpha k} = \partial \xi_\alpha / \partial x_k$ [51]. All of the integrals in (6.37) are now, at most, weakly singular if ϕ has continuous first derivatives, which we can verify by simply expanding the kernels as shown in Eqs. (6.35) and (6.36).

6.5 Discretization of the Boundary Integral Equations

We can obtain the discretized equations of the CBIE, HBIE, or Burton-Miller's BIE formulation, in either singular or weakly singular forms, by discretizing the boundary S using constant [102], linear, or quadratic [50] (see Figure 2.6) or other higher-order elements [140]. As in potential and other problems, the discretized BIEs can be written as:

$$
\begin{bmatrix} a_{11} & a_{12} & \cdots & a_{1N} \\ a_{21} & a_{22} & \cdots & a_{2N} \\ \vdots & \vdots & \ddots & \vdots \\ a_{N1} & a_{N2} & \cdots & a_{NN} \end{bmatrix} \begin{Bmatrix} \lambda_1 \\ \lambda_2 \\ \vdots \\ \lambda_N \end{Bmatrix} = \begin{Bmatrix} b_1 \\ b_2 \\ \vdots \\ b_N \end{Bmatrix}, \quad \text{or} \quad \mathbf{A}\boldsymbol{\lambda} = \mathbf{b}, \quad (6.38)
$$

where $\mathbf{A}$ is the system matrix; $\boldsymbol{\lambda}$ is the vector of unknown boundary variables at the nodes; $\mathbf{b}$ is the known vector containing contributions from the possible source term, the plane incident wave, or boundary conditions; and N is the number of nodes on the boundary. In contrast to static problems, for acoustic wave problems, this system of equations is in complex numbers; that is, all of the coefficients and variables are complex numbers, and thus the memory requirement is four times as large as its counterpart in potential problems.

As a result of this, only relatively small models can be solved by use of the conventional BEM approach with direct solvers.

6.6 Fast Multipole Boundary Element Method for 2D Acoustic Wave Problems

We first discuss the fast multipole BEM formulation for 2D acoustic wave problems (see, e.g., Ref. [41]). The 2D formulation is based on Graf's equation [see Ref. [138], p. 363, Eq. (9.1.79)] for the kernel G. That is, the far-field expansion for the G kernel can be represented as follows:

$$G(\mathbf{x}, \mathbf{y}, \omega) = \frac{i}{4} H_0^{(1)}(kr) = \frac{i}{4} \sum_{n=-\infty}^{\infty} O_n(\mathbf{x} - \mathbf{y}_c) I_{-n}(\mathbf{y} - \mathbf{y}_c), \quad |\mathbf{y} - \mathbf{y}_c| < |\mathbf{x} - \mathbf{y}_c|,$$

(6.39)

where k is the wavenumber, $\mathbf{y}_c$ is an expansion point close to $\mathbf{y}$, and the two auxiliary functions I_n and O_n are given by:

$$I_n(\mathbf{x}) = (-i)^n J_n(kr) e^{in\alpha},$$

(6.40)

$$O_n(\mathbf{x}) = i^n H_n^{(1)}(kr) e^{in\alpha}.$$

(6.41)

In the preceding two expressions, $\mathbf{x}$ is a typical vector, $J_n(\ )$ is the Bessel J function, and (r, α) is the polar coordinate of $\mathbf{x}$.

Using Eq. (6.39), we find that the far-field expansion for the F kernel is given by:

$$F(\mathbf{x}, \mathbf{y}, \omega) = \frac{\partial G(\mathbf{x}, \mathbf{y}, \omega)}{\partial n(\mathbf{y})} = \frac{i}{4} \sum_{n=-\infty}^{\infty} O_n(\mathbf{x} - \mathbf{y}_c) \frac{\partial I_{-n}(\mathbf{y} - \mathbf{y}_c)}{\partial n(\mathbf{y})},$$

$$|\mathbf{y} - \mathbf{y}_c| < |\mathbf{x} - \mathbf{y}_c|,$$

(6.42)

in which the derivative can be obtained by the formula:

$$\frac{\partial I_{-n}(\mathbf{y} - \mathbf{y}_c)}{\partial n(\mathbf{y})} = \frac{(-i)^n k}{2} [J_{n+1}(kr) e^{i\delta} - J_{n-1}(kr) e^{-i\delta}] e^{in\alpha},$$

(6.43)

where δ is the angle between the vector $\vec{r}$ from $\mathbf{y}_c$ to $\mathbf{y}$ and the outward normal $n(\mathbf{y})$.

Applying expansions in Eqs. (6.39) and (6.42), we can evaluate the G and F integrals in CBIE (6.27) on S_c (a subset of S that is away from the source point $\mathbf{x}$) with the following *multipole expansions*:

$$\int_{S_c} G(\mathbf{x}, \mathbf{y}, \omega) q(\mathbf{y}) dS(\mathbf{y}) = \sum_{n=-\infty}^{\infty} O_n(\mathbf{x} - \mathbf{y}_c) M_n(\mathbf{y}_c), \quad |\mathbf{y} - \mathbf{y}_c| < |\mathbf{x} - \mathbf{y}_c|, \quad (6.44)$$

$$\int_{S_c} F(\mathbf{x}, \mathbf{y}, \omega) \phi(\mathbf{y}) dS(\mathbf{y}) = \sum_{n=-\infty}^{\infty} O_n(\mathbf{x} - \mathbf{y}_c) \tilde{M}_n(\mathbf{y}_c), \quad |\mathbf{y} - \mathbf{y}_c| < |\mathbf{x} - \mathbf{y}_c|, \quad (6.45)$$

where M_n and $\widetilde{M}_n$ are the *multipole moments* centered at $\mathbf{y}_c$ and given by:

$$M_n(\mathbf{y}_c) = \frac{i}{4} \int_{S_c} I_{-n}(\mathbf{y} - \mathbf{y}_c) q(\mathbf{y}) dS(\mathbf{y}), \tag{6.46}$$

$$\widetilde{M}_n(\mathbf{y}_c) = \frac{i}{4} \int_{S_c} \frac{\partial I_{-n}(\mathbf{y} - \mathbf{y}_c)}{\partial n(\mathbf{y})} \phi(\mathbf{y}) dS(\mathbf{y}). \tag{6.47}$$

When the multipole expansion center is moved from $\mathbf{y}_c$ to $\mathbf{y}_{c'}$, we have the following M2M translation for both M_n and $\widetilde{M}_n$:

$$M_n(\mathbf{y}_{c'}) = \sum_{m=-\infty}^{\infty} I_{n-m}(\mathbf{y}_c - \mathbf{y}_{c'}) M_m(\mathbf{y}_c). \tag{6.48}$$

The local expansion for the G kernel integral in CBIE (6.27) is given as:

$$\int_{S_c} G(\mathbf{x}, \mathbf{y}, \omega) q(\mathbf{y}) dS(\mathbf{y}) = \sum_{n=-\infty}^{\infty} L_n(\mathbf{x}_L) I_{-n}(\mathbf{x} - \mathbf{x}_L), \tag{6.49}$$

where $\mathbf{x}_L$ is the local expansion point close to $\mathbf{x}$ $(|\mathbf{x} - \mathbf{x}_L| < |\mathbf{y}_c - \mathbf{x}_L|)$, and the expansion coefficients are given by the following M2L translation:

$$L_n(\mathbf{x}_L) = \sum_{m=-\infty}^{\infty} (-1)^m O_{n-m}(\mathbf{x}_L - \mathbf{y}_c) M_m(\mathbf{y}_c). \tag{6.50}$$

This result, which is slightly different from that given in Ref. [41], is derived with Graf's equation [138].

Similarly, the local expansion for the F kernel integral in CBIE (6.27) is given by:

$$\int_{S_c} F(\mathbf{x}, \mathbf{y}, \omega) \phi(\mathbf{y}) dS(\mathbf{y}) = \sum_{n=-\infty}^{\infty} L_n(\mathbf{x}_L) I_{-n}(\mathbf{x} - \mathbf{x}_L), \tag{6.51}$$

with $\widetilde{M}_n$ replacing M_n in M2L translation (6.50) for calculating the expansion coefficient.

The local expansion center in expansion (6.49) can be shifted from $\mathbf{x}_L$ to $\mathbf{x}_{L'}$ using the following L2L translation:

$$L_n(\mathbf{x}_{L'}) = \sum_{m=-\infty}^{\infty} I_m(\mathbf{x}_{L'} - \mathbf{x}_L) L_{n-m}(\mathbf{x}_L). \tag{6.52}$$

For HBIE (6.28), the local expansion of the K kernel integral can be written as:

$$\int_{S_c} K(\mathbf{x}, \mathbf{y}, \omega) q(\mathbf{y}) dS(\mathbf{y}) = \sum_{n=-\infty}^{\infty} L_n(\mathbf{x}_L) \frac{\partial I_{-n}(\mathbf{x} - \mathbf{x}_L)}{\partial n(\mathbf{x})}, \tag{6.53}$$

with the same local expansion coefficient $L_n(\mathbf{x}_L)$ as that given by Eq. (6.50). Similarly, the local expansion for the H kernel integral is given by:

$$\int_{S_c} H(\mathbf{x}, \mathbf{y}, \omega)\phi(\mathbf{y})dS(\mathbf{y}) = \sum_{n=-\infty}^{\infty} L_n(\mathbf{x}_L)\frac{\partial I_{-n}(\mathbf{x} - \mathbf{x}_L)}{\partial n(\mathbf{x})}, \qquad (6.54)$$

with $\tilde{M}_n$ replacing M_n in Eq. (6.50) for evaluating $L_n(\mathbf{x}_L)$. Therefore, the same moments, M2M, M2L, and L2L translations as used for the G and F integrals in the CBIE are used for the K and H integrals in the HBIE, respectively.

The fast multipole algorithms and implementations for 2D acoustic wave problems are similar to those for 2D potential problems, as given in Chapter 3. For example, the same tree structure and code for a 2D potential program can be applied to a 2D acoustic program. The only difficult part is to select a proper p, the number of expansion terms in the multipole and local expansions. For low-frequency problems, a value of p less than or equal to 10 is found to be sufficient, and for higher-frequency problems, larger values of p will be needed, and this will consume more CPU time because of the nature of the expansions for the kernels.

6.7 Fast Multipole Boundary Element Method for 3D Acoustic Wave Problems

The FMM for solving Burton–Miller's BIE (6.33) is discussed in this section for the 3D case [102]. We first note that the fundamental solution $G(\mathbf{x}, \mathbf{y}, \omega)$ for the Helmholtz equation in three dimensions can be expanded as (see, e.g., Refs. [63, 128]):

$$G(\mathbf{x},\mathbf{y}, \omega) = \frac{ik}{4\pi} \sum_{n=0}^{\infty} (2n + 1) \sum_{m=-n}^{n} O_n^m(k, \mathbf{x} - \mathbf{y}_c)\bar{I}_n^m(k, \mathbf{y} - \mathbf{y}_c),$$

$$|\mathbf{y} - \mathbf{y}_c| < |\mathbf{x} - \mathbf{y}_c|, \qquad (6.55)$$

where k is the wavenumber, $\mathbf{y}_c$ is an expansion point near $\mathbf{y}$, and the outer function O_n^m is defined by:

$$O_n^m(k, \mathbf{x}) = h_n^{(1)}(k\,|\mathbf{x}|)\, Y_n^m\left(\frac{\mathbf{x}}{|\mathbf{x}|}\right), \qquad (6.56)$$

the inner function I_n^m given by:

$$I_n^m(k, \mathbf{x}) = j_n(k\,|\mathbf{x}|)\, Y_n^m\left(\frac{\mathbf{x}}{|\mathbf{x}|}\right), \qquad (6.57)$$

and $\bar{I}_n^m$ is the complex conjugate of I_n^m. In the preceding equations, $h_n^{(1)}$ is the nth-order spherical Hankel function of the first kind, j_n is the nth-order spherical Bessel function of the first kind, and Y_n^m are the *spherical harmonics* given by:

$$Y_n^m(\mathbf{x}) = \sqrt{\frac{(n-m)!}{(n+m)!}} P_n^m(\cos\theta)e^{im\phi}, \quad \text{for } n = 0, 1, 2, \dots, \ m = -n, \dots, n,$$
(6.58)

with (ρ, θ, ϕ) being the coordinates of $\mathbf{x}$ here in a spherical coordinate system (i.e., $x_1 = \rho\sin\theta\cos\phi$, $x_2 = \rho\sin\theta\sin\phi$, $x_3 = \rho\cos\theta$), and P_n^m is the associated Legendre function defined in Eq. (3.49). These spherical harmonics are orthogonal to each other over the unit sphere and thus can form the basis for expanding other functions [136]. Note that slightly different definitions of the spherical harmonics exist in the literature [136, 137], and care needs to be taken to make sure that the fast multipole formulations are consistent with these different notations.

Similarly, the kernel $F(\mathbf{x}, \mathbf{y}, \omega)$ for 3D acoustic wave problems can be expanded as:

$$F(\mathbf{x},\mathbf{y},\omega) = \frac{ik}{4\pi}\sum_{n=0}^{\infty}(2n+1)\sum_{m=-n}^{n}O_n^m(k,\mathbf{x}-\mathbf{y}_c)\frac{\partial \bar{I}_n^m(k,\mathbf{y}-\mathbf{y}_c)}{\partial n(\mathbf{y})},$$
$$|\mathbf{y}-\mathbf{y}_c| < |\mathbf{x}-\mathbf{y}_c|. \quad (6.59)$$

Applying expansions in Eqs. (6.55) and (6.59), we can evaluate the G and F integrals in CBIE (6.27) on S_c (a subset of S that is away from the source point $\mathbf{x}$) with the following *multipole expansions*:

$$\int_{S_c}G(\mathbf{x},\mathbf{y},\omega)q(\mathbf{y})dS(\mathbf{y}) = \frac{ik}{4\pi}\sum_{n=0}^{\infty}(2n+1)\sum_{m=-n}^{n}O_n^m(k,\mathbf{x}-\mathbf{y}_c)M_{n,m}(k,\mathbf{y}_c),$$
$$|\mathbf{y}-\mathbf{y}_c| < |\mathbf{x}-\mathbf{y}_c|, \quad (6.60)$$

$$\int_{S_c}F(\mathbf{x},\mathbf{y},\omega)\phi(\mathbf{y})dS(\mathbf{y}) = \frac{ik}{4\pi}\sum_{n=0}^{\infty}(2n+1)\sum_{m=-n}^{n}O_n^m(k,\mathbf{x}-\mathbf{y}_c)\tilde{M}_{n,m}(k,\mathbf{y}_c),$$
$$|\mathbf{y}-\mathbf{y}_c| < |\mathbf{x}-\mathbf{y}_c|, \quad (6.61)$$

where $M_{n,m}$ and $\tilde{M}_{n,m}$ are the multipole moments centered at $\mathbf{y}_c$ and given by:

$$M_{n,m}(k,\mathbf{y}_c) = \int_{S_c}\bar{I}_n^m(k,\mathbf{y}-\mathbf{y}_c)q(\mathbf{y})dS(\mathbf{y}),$$
(6.62)

$$\tilde{M}_{n,m}(k,\mathbf{y}_c) = \int_{S_c}\frac{\partial \bar{I}_n^m(k,\mathbf{y}-\mathbf{y}_c)}{\partial n(\mathbf{y})}\phi(\mathbf{y})dS(\mathbf{y}).$$
(6.63)

When the multipole expansion center is moved from $\mathbf{y}_c$ to $\mathbf{y}_{c'}$, we have the following M2M translation:

$$M_{n,m}(k,\mathbf{y}_{c'}) = \sum_{n'=0}^{\infty}(2n'+1)\sum_{m'=-n'}^{n'}\sum_{\substack{l=|n-n'| \\ n'+n-l:\, \text{even}}}^{n+n'}(-1)^{m'}W_{n,n',m,m',l}$$
$$\times I_l^{-m-m'}(k,\mathbf{y}_c-\mathbf{y}_{c'})M_{n',-m'}(k,\mathbf{y}_c),$$
(6.64)

where $W_{n,n',m,m',l}$ is calculated with the following formula:

$$W_{n,n',m,m',l} = (2l+1)i^{n'-n+l} \begin{pmatrix} n & n' & l \\ 0 & 0 & 0 \end{pmatrix} \begin{pmatrix} n & n' & l \\ m & m' & -m-m' \end{pmatrix}, \quad (6.65)$$

and $\begin{pmatrix} \bullet & \bullet & \bullet \\ \bullet & \bullet & \bullet \end{pmatrix}$ denotes the Wigner $3j$ symbol [141].

The local expansion for the G kernel integral in CBIE (6.27) is given as follows:

$$\int_{S_c} G(\mathbf{x},\mathbf{y},\omega)q(\mathbf{y})dS(\mathbf{y}) = \frac{ik}{4\pi} \sum_{n=0}^{\infty} (2n+1) \sum_{m=-n}^{n} L_{n,m}(k,\mathbf{x}_L)\bar{I}_n^m(k,\mathbf{x}-\mathbf{x}_L),$$

$$(6.66)$$

where the local expansion coefficients are given by the following M2L translation:

$$L_{n,m}(k,\mathbf{x}_L) = \sum_{n'=0}^{\infty} (2n'+1) \sum_{m'=-n'}^{n'} \sum_{\substack{l=|n-n'| \\ n+n'-l:\, \text{even}}}^{n+n'} W_{n',n,m',m,l}\tilde{O}_l^{-m-m'}(k,\mathbf{x}_L-\mathbf{y}_c)$$

$$\times M_{n',m'}(k,\mathbf{y}_c), \quad (6.67)$$

for $|\mathbf{x}-\mathbf{x}_L| < |\mathbf{y}_c-\mathbf{x}_L|$, in which $\mathbf{x}_L$ is the local expansion center and $\tilde{O}_n^m$ is defined by:

$$\tilde{O}_n^m(k,\mathbf{x}) = h_n^{(1)}(k\,|\mathbf{x}|)\bar{Y}_n^m\left(\frac{\mathbf{x}}{|\mathbf{x}|}\right). \quad (6.68)$$

The local expansion center can be shifted from $\mathbf{x}_L$ to $\mathbf{x}_{L'}$ by the following L2L translation:

$$L_{n,m}(k,\mathbf{x}_{L'}) = (-1)^m \sum_{n'=0}^{\infty} (2n'+1) \sum_{m'=-n'}^{n'} \sum_{\substack{l=|n-n'| \\ n+n'-l:\, \text{even}}}^{n+n'} W_{n',n,m',-m,l}$$

$$\times I_l^{m-m'}(k,\mathbf{x}_{L'}-\mathbf{x}_L)L_{n',m'}(k,\mathbf{x}_L). \quad (6.69)$$

The local expansion for the F kernel integral in CBIE (6.27) is similar to that of Eq. (6.66):

$$\int_{S_c} F(\mathbf{x},\mathbf{y},\omega)\phi(\mathbf{y})dS(\mathbf{y}) = \frac{ik}{4\pi} \sum_{n=0}^{\infty} (2n+1) \sum_{m=-n}^{n} L_{n,m}(k,\mathbf{x}_L)\bar{I}_n^m(k,\mathbf{x}-\mathbf{x}_L),$$

$$(6.70)$$

with $\tilde{M}_{n,m}$ replacing $M_{n,m}$ in M2L translation (6.67).

For HBIE (6.28), we can obtain the local expansions for the K and H integrals by taking the normal derivatives of the local expansions for the G and F integrals, respectively. We have:

$$\int_{S_c} K(\mathbf{x},\mathbf{y},\omega)q(\mathbf{y})dS(\mathbf{y}) = \frac{ik}{4\pi} \sum_{n=0}^{\infty} (2n+1) \sum_{m=-n}^{n} L_{n,m}(k,\mathbf{x}_L)\frac{\partial \bar{I}_n^m(k,\mathbf{x}-\mathbf{x}_L)}{\partial n(\mathbf{x})},$$

$$(6.71)$$

with $M_{n,m}$ in M2L translation (6.67), and similarly for the H kernel integral:

$$\int_{S_c} H(\mathbf{x}, \mathbf{y}, \omega)\phi(\mathbf{y})dS(\mathbf{y}) = \frac{ik}{4\pi} \sum_{n=0}^{\infty} (2n+1) \sum_{m=-n}^{n} L_{n,m}(k, \mathbf{x}_L) \frac{\partial \bar{I}_n^m(k, \mathbf{x} - \mathbf{x}_L)}{\partial n(\mathbf{x})},$$

(6.72)

with $\widetilde{M}_{n,m}$ replacing $M_{n,m}$ in M2L translation (6.67). Again, the same moments, M2M, M2L, and L2L translations, as used for the G and F integrals in the CBIE are used for the K and H integrals in the HBIE, respectively.

To determine p, the order of the multipole and local expansions, the following empirical formula (see, e.g., Ref. [41]) can be applied:

$$p = kD + c_0 \log(kD + \pi),$$

(6.73)

where D is the diameter of the cell on which the expansions are calculated and c_0 is a number that depends on the precision of the arithmetic. Formulas like (6.73) can be applied to adaptively determine the values of p at different tree levels in the fast multipole algorithms.

The fast multipole formulations just discussed for solving 3D acoustic wave problems or Helmholtz equations in general are good for low frequencies because of the $O(p^5)$ nature of the formulation. To perform the M2M, M2L, and L2L translations, $O(p^5)$ computations are required because there are three summations in all of these translations and two indices in the coefficients, as shown in Eqs. (6.64), (6.67), and (6.69). Although the number of operations can be reduced to $O(p^4)$ by use of various recursive relations, the computing time can still increase quickly with the increase of the value of p. In addition, the use of the Wigner $3j$ symbol in Eq. (6.65), which is time-consuming to calculate each time and consumes more memory if its values are stored, further reduces the computational efficiency. As mentioned in the 2D case and also shown in Eq. (6.73), at higher frequencies, more terms are required in the expansions to represent the increased variations in the field, leading to a larger p and a slower performance of any fast multipole BEM code based on the formulations discussed previously. In fact, the FMM gives $O(N^2)$ computing complexity using these original formulations [41]. Adaptive fast multipole algorithms [60, 61] are used to accelerate the solutions of the fast multipole BEM for 3D acoustic wave problems based on these formulations [102]. Large acoustic BEM models with total DOFs (in complex variables) of up to 200,000 are solved at lower frequencies on a laptop with only 512-MB RAM [102].

For higher-frequency problems, the diagonal form proposed by Rokhlin [127] can be used to accelerate the computations of all the translations. Unfortunately, this diagonal form breaks down at lower frequencies, where the original formulations will need to be applied [41]. The wideband FMM proposed by Cheng *et al.* [137] may be considered; it provides a seamless framework for combining the low- and high-frequency formulations. Conversely, an $O(p^3)$

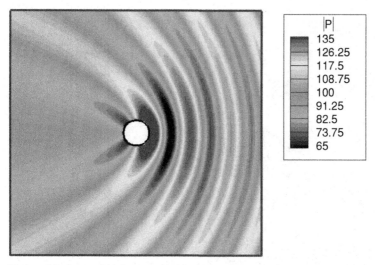

The figure legend shows:

|P|
135
126.25
117.5
108.75
100
91.25
82.5
73.75
65

Figure 6.3. Scattering from a single cylinder with 1000 elements.

formulation was developed by Gumerov and Duraiswami [132, 136] that is adequate for both low- and high-frequency applications. This $O(p^3)$ formulation does not use the Wigner $3j$ symbol, which also can reduce the memory usage.

Based on the adaptive fast multipole algorithms reported in Ref. [102] and the Gumerov and Duraiswami's $O(p^3)$ formulations presented in Ref. [136], a very robust acoustic software, *FastBEM Acoustics*®, was developed that has been applied successfully in solving large-scale acoustic BEM models with DOFs above 2 million on desktop PCs with 32-bit operating systems. Several 3D numerical examples presented in the following section are solved with the *FastBEM Acoustics*® (V.1.2.0) software.

6.8 Numerical Examples

Several examples of 2D and 3D acoustic wave problems are presented in this section. Constant elements are used in all cases; that is, one-node line elements for 2D models and one-node triangular elements for 3D problems. The 2D computer code is based on the formulations presented in Section 6.6 and the same fast multipole algorithms for 2D potential problems discussed in Chapter 3. The 3D code, named *FastBEM Acoustics*®, is based on an improved adaptive fast multipole algorithm of that in Ref. [102] and the $O(p^3)$ formulations in Ref. [136].

6.8.1 Scattering from Cylinders in a 2D Medium

First, we consider a 2D scattering problem with a single rigid cylinder with the incident wave coming from the right. The cylinder has a radius $a = 1$ and is discretized with 1000 elements. A relative error of 0.01 % is achieved for $ka =$

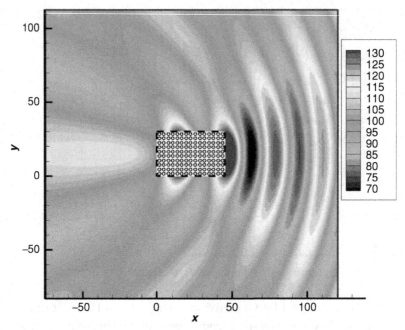

Figure 6.4. Scattering from multiple cylinders with 195,000 elements.

1 with 16 expansion terms and a tolerance of 10^{-8}. Figure 6.3 is the contour plot of the magnitude of the scattered-pressure field outside the cylinder in a square-field region.

We next consider a case with multiple scatterers in which an array of 15×10 cylinders is modeled with 195,000 elements. Again, the incident wave is from the right. Figure 6.4 shows the computed scattered-pressure field with 10,000 field points inside the domain and at $ka = 5.4$. The model was solved in about 3.5 h for a tolerance of 10^{-8} on an Intel® Core2 Duo desktop PC. In this case, about 2 h of the total CPU time was spent on solving the BEM system of equations and another 1.5 h on calculating the pressure at the 10,000 field points inside the domain, which also can be accelerated by the fast multipole algorithms.

6.8.2 Radiation from a Pulsating Sphere

A pulsating sphere with radius $a = 1$ m is used next to verify the 3D fast multipole BEM code. The analytical solution for this problem was discussed in Section 6.1. We consider the case with changing frequencies (frequency sweep) for the nondimensional wavenumber ka varying from 0.1 to 10. The total number of elements used is 1200. The computed pressures at $(5a, 0, 0)$ are plotted in Figure 6.5, which shows that the conventional BEM with the CBIE fails to

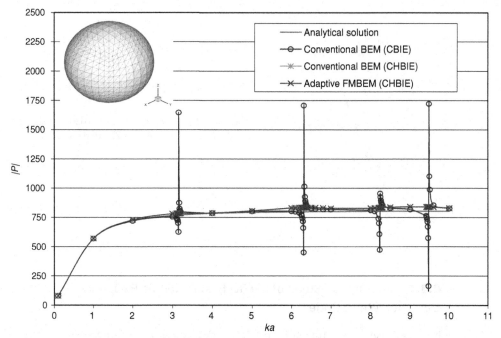

Figure 6.5. Frequency-sweep plot for the pulsating-sphere model.

predict the surface-pressure field at the fictitious eigenfrequencies ($ka = \pi$, $2\pi,\ldots$, for this case). The results obtained with the conventional BEM with the Burton–Miller (CHBIE) formulation agree well with the analytical solution at all wavenumbers. The adaptive fast multipole BEM with the CHBIE also yields results very close to those of the conventional BEM with the CHBIE, suggesting that the truncation errors introduced in multipole expansions are under control for ka ranging from 0.1 to 10. In this example, the maximum number of elements in a leaf is set to 100, the number of multipole and local expansion terms to 10, and the tolerance for GMRES to 10^{-3}.

6.8.3 Scattering from Multiple Scatterers

A multiscatterer model (Figure 6.6) containing 1000 randomly distributed capsulelike rigid scatterers in a 2 m × 2 m × 2 m domain is studied next. Each scatterer is meshed with 200 elements, with a total of 200,000 elements for the entire model. The incident wave is e^{-ikx} with $k = 1$. Sample field points are taken at an annular field surface with inner and outer radii equal to 5 m and 10 m, respectively. The computed sound-pressure distribution is shown in Figure 6.7 for this discretization. The total CPU time used to solve this large model is 56 min on a HP laptop with an Intel® 1.6-GHz Centrino CPU and 512-MB RAM.

Figure 6.6. A BEM model of 1000 capsulelike scatterers with a total of 200,000 elements.

6.8.4 Performance Study of the 3D Fast Multipole Boundary Element Method Code

Next, we use a radiating sphere to test the accuracy and efficiency of the 3D acoustic fast multipole BEM code. The numbers of elements used range from 588 to 1,503,792. The nondimensional wavenumbers are $ka = 2$ and 20, with corresponding initial numbers of expansion terms $p = 6$ and 10, respectively. All the BEM models were solved on a Dell® PC with Intel® Core2 Duo CPU at 2.2 GHz and 2-GB RAM. The tolerance for convergence is set at 10^{-4}.

Figure 6.8 shows the relative errors in the computed sound pressure and power (integration of the pressure and velocity on the surface). As seen from

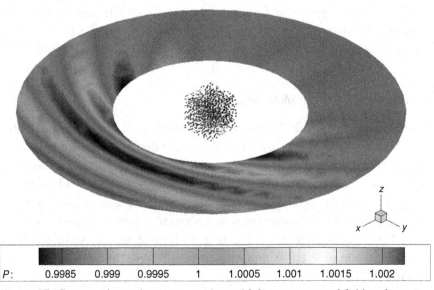

| P: | 0.9985 | 0.999 | 0.9995 | 1 | 1.0005 | 1.001 | 1.0015 | 1.002 |

Figure 6.7. Computed sound pressure on the multiple scatterers and field surface.

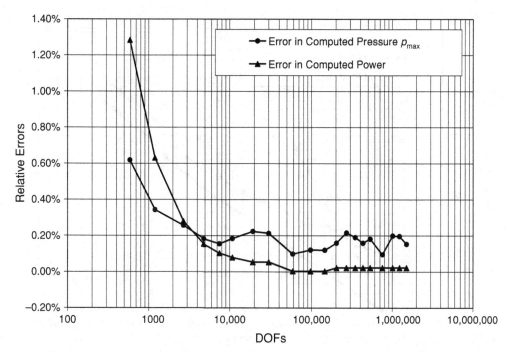

Figure 6.8. Relative errors of the fast multipole BEM solutions for a pulsating sphere at $ka = 2$.

the plot, the accuracy of the fast multipole BEM is quite satisfactory considering the tolerance (10^{-4}) used and the sizes of the BEM models. The errors decrease quickly and stay around 0.2% for models with more than 100,000 DOFs at $ka = 2$, indicating the numerical stability of the used fast multipole algorithms.

Figure 6.9 shows the CPU time used by the fast multipole BEM code compared with that of a conventional BEM code. The CPU time for the fast multipole BEM code increases almost linearly with the increase of the DOFs, and the largest BEM model with 1.5 million DOFs was solved within 65 min at $ka = 2$ (note that this is a system of equations with complex variables, which is equivalent to a system of about 3 million DOFs in real variables). The conventional BEM, however, can only solve models with up to 10,800 DOFs on the same PC, and the CPU time used increases almost as a cubic function of the DOFs. The efficiencies of the fast multipole BEM compared with those of the conventional BEM are most evident from this example.

6.8.5 An Engine-Block Model

We next study the radiation of acoustic waves from an engine block. The engine block has an overall dimension of 0.31 m $\times 0.27$ m $\times 0.36$ m in the

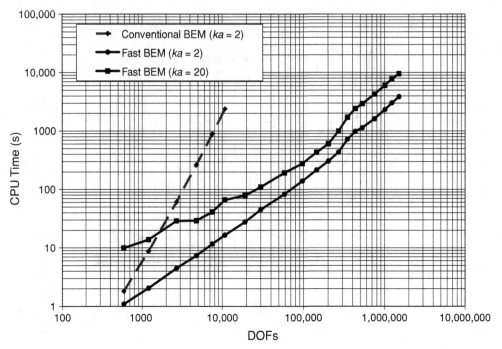

Figure 6.9. Total CPU time used to solve the pulsating-sphere model.

x, y, and z directions, respectively, and is discretized with 51,766 triangular elements, as shown in Figure 6.10. Velocity BCs are applied on the surfaces of the engine block, and the sound-pressure field on a semispherical field surface is sought at wavenumber $ka = 3.6$. Figure 6.11 shows the computed

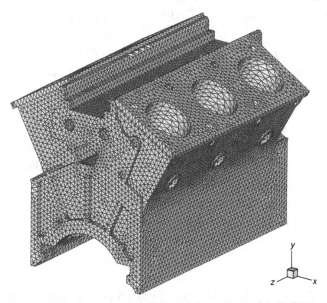

Figure 6.10. An engine-block model discretized with 51,766 boundary elements.

Figure 6.11. Computed sound-pressure distribution for the engine-block model.

sound-pressure distribution on this field surface. The model was solved in 9 min on a Dell® PC with Intel® Core2 Duo CPU and with the tolerance set at 10^{-4}.

6.8.6 A Submarine Model

A submarine model is studied next to predict the noise that is due to the vibration of the propeller. This is an interesting example of using the fast multipole BEM in solving large-scale underwater acoustic problems, which has been a challenging task for other domain-based methods. The Skipjack submarine is modeled, which has a length of 76.8 m. A total of 250,220 elements are used, with a typical element size equal to 0.14 m. Velocity BCs are applied to the propeller, and the model is solved at a nondimensional wavenumber $ka = 38.4$ (frequency $f = 123.3$ Hz). The computed sound-pressure level on the surface of the submarine is shown in Figure 6.12, and the radiated wave on a cylindrical field surface is shown in Figure 6.13. The model was solved in 54 min on a Dell® PC with Intel® Core2 Duo CPU and with the tolerance set at 10^{-4}. Scattering problems, in which the submarine is motionless and incident waves impinge on the model from different directions, were also solved with the same BEM model.

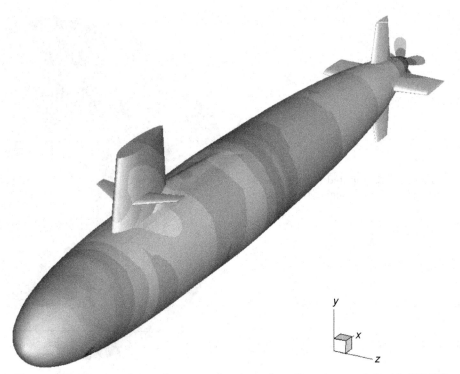

Figure 6.12. Computed sound-pressure level on the submarine model with 250,220 elements.

6.8.7 An Airbus A320 Model

In this example, prediction of the jet noise from an airplane is attempted by using the fast multipole BEM code. A model of Airbus A320 is used, which has a length of 123 ft. There are 541,152 elements for this model with a typical element size equal to 0.2 ft. The plot of the BEM mesh on one of the engines is shown in Figure 6.14. The acoustic pressure on the surface of the airplane that is due to the vibrations of the two jet engines was computed at $ka = 12.3$ and is shown in Figure 6.15. The model was solved in 131 min on a Dell® PC with a tolerance of 10^{-4}. Prediction of jet noise is still a very challenging problem for the fast multipole BEM because of the nature of the high frequencies involved. Large BEM models with considerably more elements need to be used, and the models need to be solved on PC clusters or supercomputers.

6.8.8 A Human-Head Model

A human-head model is now presented for acoustic analysis. Such models using the BEM can be used to study the impact of noise on human hearing and to help design better audio devices. The head model is discretized using 87,340

Figure 6.13. Radiated sound pressure on the field surface.

elements, and the sound pressure on the surface of the model for a plane incident wave coming in the $-x$ direction and at 11 kHz ($ka = 50$) is computed, as shown in Figure 6.16. The model was solved in 10 min on a Dell® PC with a tolerance of 10^{-4}. It is interesting to note that both ears on the illuminated

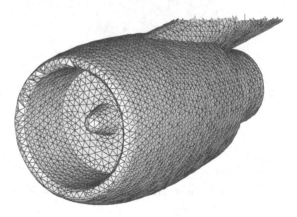

Figure 6.14. Plot of BEM mesh near one engine of the Airbus A320 model (with a total of 541,152 elements).

Figure 6.15. Sound-pressure distribution on the Airbus A320 model.

side (left ear) and the shadow side (right ear) register higher values of sound pressure, in addition to the area between the two lips, indicating the unique acoustic effects of the geometries near these areas. The same phenomenon is observed in models at other frequencies or with the sound from other directions.

6.8.9 Analysis of Sound Barriers – A Half-Space Acoustic Wave Problem

Many of the acoustic problems are present in a half-space, such as airport or other traffic noise control problems. With the BEM, these half-space acoustic wave problems also can be modeled readily. Formulations of the adaptive fast multipole BEM for 3D half-space acoustic wave problems can be found in Ref. [142].

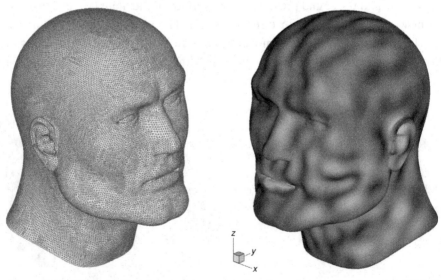

Figure 6.16. BEM mesh and sound-pressure plots for a human-head model (87,340 elements).

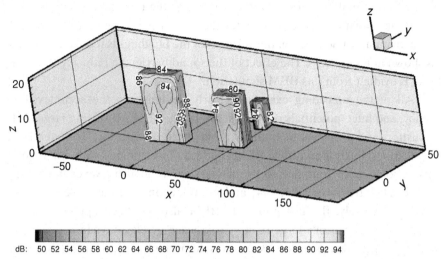

Figure 6.17. Noise level (in decibels) on buildings without a barrier.

Figures 6.17 and 6.18 show the computed sound-pressure levels (in decibels) for the BEM models of three buildings near a highway without and with a sound barrier, respectively, using the fast multipole BEM for half-space acoustic wave problems [142]. The dimensions (length × width × height) of the three buildings are 30 × 10 × 20, 20 × 12 × 15, and 9.5 × 9 × 8 (in meters), respectively. The barrier has a height of 6 m and a length of 255.94 m. One point source load with a frequency of 20 Hz is located 13 m away from the middle point of the barrier and 1 m above the ground. The BEM model contains 56,465 triangular elements. In the case with no sound barrier, the surface

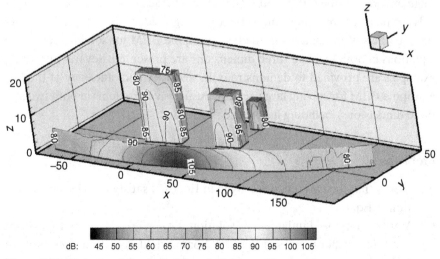

Figure 6.18. Noise level (in decibels) on buildings with a barrier.

of the largest building closest to the source has the maximum sound level of 94 dB, as shown in Figure 6.17. After the barrier is inserted into the model, the maximum sound level on the surfaces of the buildings is reduced to 90 dB, as shown in Figure 6.18. The effect of the sound barriers in reducing the noise level is evident from this BEM simulation.

All of the preceding examples clearly demonstrate the accuracy, efficiency, and huge potentials of the fast multipole BEM for solving large-scale acoustic wave problems in both two and three dimensions. To extend the applications, the fast multipole BEM codes can be combined with other methods to solve more complicated problems, such as acoustic waves interacting with elastic structures [143, 144] and multidomain acoustic wave problems [145]. Conversely, the fast multipole BEM also has been applied successfully in solving various large-scale *elastic wave* or elastodynamic problems, and extensive research results on this important topic can be found in Refs. [41, 63].

6.9 Summary

The basic governing equations for acoustic wave problems are reviewed in this chapter. The main equation to be solved in acoustic wave problems is the Helmholtz equation, which reduces to the Laplace equation for potential problems if the wavenumber is zero. Thus, the acoustic wave problems are closely related to the potential problems we studied in Chapters 2 and 3. The fundamental solution for the Helmholtz equation is derived for 3D cases using the solution of a pulsating sphere. BIE formulations are presented with the emphasis on the Burton–Miller BIE formulation, which can provide unique solutions for all wavenumbers for exterior acoustic wave problems. Weakly singular forms of the CBIE and HBIE are introduced by using the static kernels for potential problems and using the integral identities satisfied by these static kernels. Formulations in the fast multipole BEM for solving Helmholtz equations in both two and three dimensions are presented. Several numerical examples are provided to demonstrate the accuracy and efficiency of the fast multipole BEM for solving large-scale acoustic wave problems in both two and three dimensions, including half-space problems.

Problems

6.1. Verify that the two functions given in Eq. (6.2) satisfy the 1D wave equation in Eq. (6.1).

6.2. Verify that the fundamental solution $G(\mathbf{x}, \mathbf{y}, \omega)$ given in Eq. (6.18) for three dimensions satisfies the Sommerfeld radiation condition in Eq. (6.11).

6.3. Derive the kernels K and H in Eqs. (6.29) and (6.30) for the 3D HBIE.

6.4. Prove Eq. (6.36); that is, show that $r_{,n} = \frac{\partial r}{\partial n(\mathbf{y})} = O(r)$ as $r \to 0$.

6.5. Show that all the integrals in Eq. (6.38) are weakly singular; that is, all of the integrands have $O(1/r)$ or less singularity in 3D cases.

6.6. Because the constant solution or rigid-body solution approach does not apply to the Helmholtz equation, discuss how we can determine the diagonal coefficients in the matrix associated with the F kernel for acoustic wave problems.

6.7. Write a 2D acoustic conventional BEM code using the CBIE and constant elements, based on the code for 2D potential problems in Appendix B.1. Test your code on a "pulsating-cylinder" problem.

6.8. Write a 2D acoustic fast multipole BEM code using the CBIE and constant elements, based on the code for 2D potential problems in Appendix B.2. Test your code on the "pulsating-cylinder" problem and study its accuracy and efficiency.

Analytical Integration of the Kernels

A.1 2D Potential Boundary Integral Equations

For 2D potential problems, we have the following four kernels for the CBIE and HBIE:

$$G(\mathbf{x}, \mathbf{y}) = \frac{1}{2\pi} \log\left(\frac{1}{r}\right), \tag{A.1}$$

$$F(\mathbf{x}, \mathbf{y}) = \frac{\partial G(\mathbf{x}, \mathbf{y})}{\partial n(\mathbf{y})} = -\frac{1}{2\pi r} r_{,k}\, n_k(\mathbf{y}), \tag{A.2}$$

$$K(\mathbf{x}, \mathbf{y}) = \frac{\partial G(\mathbf{x}, \mathbf{y})}{\partial n(\mathbf{x})} = \frac{1}{2\pi r} r_{,k}\, n_k(\mathbf{x}), \tag{A.3}$$

$$H(\mathbf{x}, \mathbf{y}) = \frac{\partial^2 G(\mathbf{x}, \mathbf{y})}{\partial n(\mathbf{x})\partial n(\mathbf{y})} = \frac{1}{2\pi r^2} [n_k(\mathbf{x})n_k(\mathbf{y}) - 2r_{,k}\, n_k(\mathbf{x})r_{,l}\, n_l(\mathbf{y})]. \tag{A.4}$$

The integrations of the four kernels on a line segment ΔS shown in Figure A.1 (from point 1 to point 2) can be evaluated analytically as follows (note that on ΔS, $r = d/\cos\theta$, $dS = r\,d\theta/\cos\theta$):

$$\int_{\Delta S} G(\mathbf{x}, \mathbf{y})dS = \frac{1}{2\pi} [-(\theta_2 - \theta_1)d + 2R - T_2 \log r_2 + T_1 \log r_1], \tag{A.5}$$

$$\int_{\Delta S} F(\mathbf{x}, \mathbf{y})dS = -\frac{1}{2\pi}(\theta_2 - \theta_1), \tag{A.6}$$

$$\int_{\Delta S} K(\mathbf{x}, \mathbf{y})dS = \frac{1}{2\pi}\left[(\theta_2 - \theta_1)n_k(\mathbf{y}) + \log\left(\frac{r_2}{r_1}\right)t_k(\mathbf{y})\right]n_k(\mathbf{x}), \tag{A.7}$$

$$\int_{\Delta S} H(\mathbf{x}, \mathbf{y})dS = \frac{1}{2\pi}\left[-\left(\frac{T_2}{r_2^2} - \frac{T_1}{r_1^2}\right)n_k(\mathbf{y}) + d\left(\frac{1}{r_2^2} - \frac{1}{r_1^2}\right)t_k(\mathbf{y})\right]n_k(\mathbf{x}),$$
$$\tag{A.8}$$

in which $2R(= T_2 - T_1)$ is the total length of the line element and t_k is the component of the tangential vector t (Figure A.1). These results can be used to evaluate directly the coefficients of the CBIE and HBIE for 2D potential

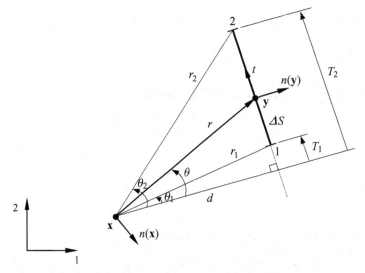

Figure A.1. Analytical integration on an arbitrary line segment.

problems using constant elements. If the source point $\mathbf{x}$ is on the element of integration (at the midpoint), we have:

$$\theta_2 - \theta_1 = \pi, \ d = 0, \ r_1 = r_2 = R, \ T_1 = -T_2 = -R,$$

and the four integrals have the following values:

$$\int_{\Delta S} G(\mathbf{x}, \mathbf{y})dS = \frac{R}{\pi}(1 - \log R),\tag{A.9}$$

$$\int_{\Delta S} F(\mathbf{x}, \mathbf{y})dS = -\frac{1}{2},\tag{A.10}$$

$$\int_{\Delta S} K(\mathbf{x}, \mathbf{y})dS = \frac{1}{2},\tag{A.11}$$

$$\int_{\Delta S} H(\mathbf{x}, \mathbf{y})dS = -\frac{1}{\pi R}.\tag{A.12}$$

Note that in the preceding results, the second (F) and third (K) integrals are equal to the CPV integrals plus the jump terms, and the last (H) integral is a HFP integral.

A.2 2D Elastostatic Boundary Integral Equations

For 2D elasticity, we have the following four kernels for the CBIE and HBIE:

$$U_{ij}(\mathbf{x}, \mathbf{y}) = \frac{1}{8\pi\mu(1-v)}\left[(3-4v)\delta_{ij}\log\left(\frac{1}{r}\right) + r_{,i}\,r_{,j} - \frac{1}{2}\delta_{ij}\right],\tag{A.13}$$

$$T_{ij}(\mathbf{x}, \mathbf{y}) = -\frac{1}{4\pi(1-v)r} \left\{ \frac{\partial r}{\partial n} \left[(1-2v)\delta_{ij} + 2r_{,i}\, r_{,j} \right] - (1-2v)\left(r_{,i}\, n_j - r_{,j}\, n_i \right) \right\},$$

$$(A.14)$$

$$K_{ij}(\mathbf{x}, \mathbf{y}) = \frac{1}{4\pi(1-v)r} \left[(1-2v)(\delta_{ij}r_{,k} + \delta_{jk}r_{,i} - \delta_{ik}r_{,j}) + 2r_{,i}\, r_{,j}\, r_{,k} \right] n_k(\mathbf{x}),$$

$$(A.15)$$

$$H_{ij}(\mathbf{x}, \mathbf{y}) = \frac{\mu}{2\pi(1-v)r^2} \left\{ 2\frac{\partial r}{\partial n} \left[(1-2v)\delta_{ik}r_{,j} + v(\delta_{ij}r_{,k} + \delta_{jk}r_{,i}) - 4r_{,i}\, r_{,j}\, r_{,k} \right] \right.$$

$$+ 2v(n_i r_{,j}\, r_{,k} + n_k r_{,i}\, r_{,j}) - (1-4v)\delta_{ik}n_j$$

$$\left. + (1-2v)(2n_j r_{,i}\, r_{,k} + \delta_{ij}n_k + \delta_{jk}n_i) \right\} n_k(\mathbf{x}).$$

$$(A.16)$$

To evaluate the integrals of these kernels over the straight-line segment ΔS (a constant element) shown in Figure A.1, we use the local coordinate system n–t at $\mathbf{y}$ on ΔS. In this local coordinate system, we have:

$$\int_{\Delta S} U_{ij}^{(n-t)}(\mathbf{x}, \mathbf{y})\, dS = \frac{1}{8\pi\mu(1-v)} \left[(3-4v)I_0\delta_{ij} + I_{ij}^{(n-t)} - R\delta_{ij} \right], \qquad (A.17)$$

$$\int_{\Delta S} T_{ij}^{(n-t)}(\mathbf{x}, \mathbf{y})\, dS = -\frac{1}{4\pi(1-v)} \left[(1-2v)\left(\Omega_0\delta_{ij} - \Omega_i^{(n-t)}\delta_{1j} + \Omega_j^{(n-t)}\delta_{1i} \right) \right.$$

$$\left. + 2\Omega_{ij}^{(n-t)} \right], \qquad (A.18)$$

$$\int_{\Delta S} K_{ij}^{(n-t)}(\mathbf{x}, \mathbf{y})\, dS = C_{ijk}n_k(\mathbf{x}), \qquad (A.19)$$

$$\int_{\Delta S} H_{ij}^{(n-t)}(\mathbf{x}, \mathbf{y})\, dS = D_{ijk}n_k(\mathbf{x}), \qquad (A.20)$$

where:

$$4\pi(1-v)C_{111} = (1-2v)\Omega_1^{(n-t)} + 2\Omega_{11}^{(n-t)},$$

$$4\pi(1-v)C_{112} = (1-2v)\Omega_2^{(n-t)} + 2\Omega_{12}^{(n-t)},$$

$$4\pi(1-v)C_{121} = -(1-2v)\Omega_2^{(n-t)} + 2\Omega_{12}^{(n-t)},$$

$$4\pi(1-v)C_{122} = (1-2v)\Omega_1^{(n-t)} + 2\Omega_{22}^{(n-t)}, \qquad (A.21)$$

$$4\pi(1-v)C_{212} = -(1-2v)\Omega_1^{(n-t)} + 2\Omega_{22}^{(n-t)},$$

$$4\pi(1-v)C_{222} = (1-2v)\Omega_2^{(n-t)} + 2\Omega_{222}^{(n-t)},$$

$$C_{211} = C_{112},\ C_{221} = C_{122};$$

$$\frac{2\pi(1-v)}{\mu}D_{111} = -\Gamma_{12} - 2d^2\Gamma_{14},$$

$$\frac{2\pi(1-v)}{\mu}D_{112} = -d\Gamma_{02} + 2d^3\Gamma_{04},$$

$$\frac{2\pi(1-v)}{\mu}D_{122} = -\Gamma_{12} + 2d^2\Gamma_{14},$$ (A.22)

$$\frac{2\pi(1-v)}{\mu}D_{222} = 3d\Gamma_{02} - 2d^3\Gamma_{04},$$

$$D_{211} = D_{121} = D_{112}, \quad D_{212} = D_{221} = D_{122}.$$

In the preceding expressions:

$$I_0 = -d(\theta_2 - \theta_1) + 2R - T_2\log r_2 + T_1\log r_1,$$

$$I_{11}^{(n-t)} = d(\theta_2 - \theta_1),$$

$$I_{12}^{(n-t)} = I_{21}^{(n-t)} = d\log(r_2/r_1),$$ (A.23)

$$I_{22}^{(n-t)} = T_2 - T_1 - d(\theta_2 - \theta_1);$$

$$\Omega_0 = \theta_2 - \theta_1,$$

$$\Omega_1^{(n-t)} = \theta_2 - \theta_1,$$

$$\Omega_2^{(n-t)} = \log(r_2/r_1),$$

$$\Omega_{11}^{(n-t)} = (\theta_2 - \theta_1)/2 + d\Gamma_{12}/2,$$ (A.24)

$$\Omega_{12}^{(n-t)} = \Omega_{21}^{(n-t)} = \Gamma_{22}/2,$$

$$\Omega_{22}^{(n-t)} = (\theta_2 - \theta_1)/2 - d\Gamma_{12}/2,$$

$$\Omega_{222}^{(n-t)} = \log(r_2/r_1) + d^2\Gamma_{02}/2;$$

$$\Gamma_{02} = 1/r_2^2 - 1/r_1^2,$$

$$\Gamma_{04} = 1/r_2^4 - 1/r_1^4,$$

$$\Gamma_{12} = T_2/r_2^2 - T_1/r_1^2,$$ (A.25)

$$\Gamma_{14} = T_2/r_2^4 - T_1/r_1^4,$$

$$\Gamma_{22} = T_2^2/r_2^2 - T_1^2/r_1^2;$$

in which all the parameters are as defined in Figure A.1. Once the integrals of the kernels are determined in the local n–t system, they need to be transformed to the global x–y system.

When the source point $\mathbf{x}$ is on the element of integration, we have:

$$\theta_2 - \theta_1 = \pi, \ d = 0, \ r_1 = r_2 = R, \ T_1 = -T_2 = -R,$$

and the four integrals have the following results:

$$\int_{\Delta S} U_{11}^{(n-t)}(\mathbf{x}, \mathbf{y})dS = \frac{R}{8\pi\mu(1-\nu)} [2(3-4\nu)(1-\log R) - 1],$$

$$\int_{\Delta S} U_{12}^{(n-t)}(\mathbf{x}, \mathbf{y})dS = \int_{\Delta S} U_{21}^{(n-t)}(\mathbf{x}, \mathbf{y})dS = 0, \tag{A.26}$$

$$\int_{\Delta S} U_{22}^{(n-t)}(\mathbf{x}, \mathbf{y})dS = \frac{R}{8\pi\mu(1-\nu)} [2(3-4\nu)(1-\log R) + 1],$$

$$\int_{\Delta S} T_{ij}^{(n-t)}(\mathbf{x}, \mathbf{y})dS = -\frac{1}{2}\delta_{ij}, \tag{A.27}$$

$$\int_{\Delta S} K_{ij}^{(n-t)}(\mathbf{x}, \mathbf{y})dS = \frac{1}{2}\delta_{ij}, \tag{A.28}$$

$$\int_{\Delta S} H_{ij}^{(n-t)}(\mathbf{x}, \mathbf{y})dS = -\frac{\mu}{\pi(1-\nu)R}\delta_{ij}. \tag{A.29}$$

Similar to the potential case, the second (T) and third (K) integrals are equal to the CPV integrals plus the jump terms, and the last (H) integral is a HFP integral.

A.3 2D Stokes Flow Boundary Integral Equations

For 2D Stokes flow problems, we have the following four kernels for the CBIE and the HBIE:

$$U_{ij}(\mathbf{x}, \mathbf{y}) = \frac{1}{4\pi\mu} \left[\delta_{ij} \log\left(\frac{1}{r}\right) + r_{,i} r_{,j} - \frac{1}{2}\delta_{ij} \right], \tag{A.30}$$

$$T_{ij}(\mathbf{x}, \mathbf{y}) = -\frac{1}{\pi r} r_{,i} r_{,j} r_{,k} n_k(\mathbf{y}), \tag{A.31}$$

$$K_{ij}(\mathbf{x}, \mathbf{y}) = \frac{1}{\pi r} r_{,i} r_{,j} r_{,k} n_k(\mathbf{x}), \tag{A.32}$$

$$H_{ij}(\mathbf{x}, \mathbf{y}) = \frac{\mu}{\pi r^2} [(\delta_{ij} r_{,k} + \delta_{jk} r_{,i} - 8r_{,i} r_{,j} r_{,k}) r_{,l} n_l(\mathbf{y})$$

$$+ n_i r_{,j} r_{,k} + n_k r_{,i} r_{,j} + \delta_{ik} n_j] n_k(\mathbf{x}). \tag{A.33}$$

The integrals of these kernels over the straight-line segment ΔS shown in Figure A.1 can be obtained from the results for 2D elasticity problems by setting Poisson's ratio $v = \frac{1}{2}$. In the local coordinate system, we obtain (see results in the previous section):

$$\int_{\Delta S} U_{ij}^{(n-t)}(\mathbf{x}, \mathbf{y})dS = \frac{1}{4\pi\mu}\left[I_0\delta_{ij} + I_{ij}^{(n-t)} - R\delta_{ij}\right], \tag{A.34}$$

$$\int_{\Delta S} T_{ij}^{(n-t)}(\mathbf{x}, \mathbf{y})dS = -\frac{1}{\pi}\Omega_{ij}^{(n-t)}, \tag{A.35}$$

$$\int_{\Delta S} K_{ij}^{(n-t)}(\mathbf{x}, \mathbf{y})dS = C_{ijk}n_k(\mathbf{x}), \tag{A.36}$$

$$\int_{\Delta S} H_{ij}^{(n-t)}(\mathbf{x}, \mathbf{y})dS = D_{ijk}n_k(\mathbf{x}), \tag{A.37}$$

where:

$$2\pi C_{111} = 2\Omega_{11}^{(n-t)},$$

$$2\pi C_{112} = 2\Omega_{12}^{(n-t)},$$

$$2\pi C_{122} = 2\Omega_{22}^{(n-t)}, \tag{A.38}$$

$$2\pi C_{222} = 2\Omega_{222}^{(n-t)},$$

$$C_{121} = C_{211} = C_{112}, \quad C_{212} = C_{221} = C_{122},$$

$$\frac{\pi}{\mu}D_{111} = -\Gamma_{12} - 2d^2\Gamma_{14},$$

$$\frac{\pi}{\mu}D_{112} = -d\Gamma_{02} + 2d^3\Gamma_{04},$$

$$\frac{\pi}{\mu}D_{122} = -\Gamma_{12} + 2d^2\Gamma_{14}, \tag{A.39}$$

$$\frac{\pi}{\mu}D_{222} = 3d\Gamma_{02} - 2d^3\Gamma_{04},$$

$$D_{211} = D_{121} = D_{112}, \quad D_{212} = D_{221} = D_{122},$$

and all the parameters I, Ω, and Γ are as defined earlier for elasticity kernels in Eqs. (A.23)–(A.25). Once the integrals of the kernels are determined in the local n–t system, they need to be transformed to the global x–y system.

When the source point $\mathbf{x}$ is on the element of integration, we have the following results for the four integrals:

$$\int_{\Delta S} U_{11}^{(n-t)}(\mathbf{x}, \mathbf{y})dS = \frac{R}{4\pi\mu}\left[2(1 - \log R) - 1\right],$$

$$\int_{\Delta S} U_{12}^{(n-t)}(\mathbf{x}, \mathbf{y})dS = \int_{\Delta S} U_{21}^{(n-t)}(\mathbf{x}, \mathbf{y})dS = 0, \qquad (A.40)$$

$$\int_{\Delta S} U_{22}^{(n-t)}(\mathbf{x}, \mathbf{y})dS = \frac{R}{4\pi\mu}\left[2(1 - \log R) + 1\right],$$

$$\int_{\Delta S} T_{ij}^{(n-t)}(\mathbf{x}, \mathbf{y})dS = -\frac{1}{2}\delta_{ij}, \qquad (A.41)$$

$$\int_{\Delta S} K_{ij}^{(n-t)}(\mathbf{x}, \mathbf{y})dS = \frac{1}{2}\delta_{ij}, \qquad (A.42)$$

$$\int_{\Delta S} H_{ij}^{(n-t)}(\mathbf{x}, \mathbf{y})dS = -\frac{2\mu}{\pi R}\delta_{ij}. \qquad (A.43)$$

Similar to the potential and elasticity cases, the second (T) and third (K) integrals are equal to the CPV integrals plus the jump terms, whereas the last (H) integral is a HFP integral.

Sample Computer Programs

B.1 A Fortran Code of the Conventional Boundary Element Method for 2D Potential Problems

The following is a list of the source code written in Fortran for the program discussed in Section 2.11 for 2D potential problems using the conventional BEM. The direct solver *dgesv* from LAPACK can be downloaded from the website www.netlib.org.

```
c----------------------------------------------------------------------
c Program:   2D_Potential - A boundary element method (BEM) code in Fortran
c            for analyzing general 2D potential problems (governed by
c            Laplace equation) using constant elements.
c
c Developer: Dr. Yijun Liu at the University of Cincinnati, Cincinnati, OH.
c
c Version:   V.1.20.
c Released:  October 1, 2008.
c
c Copyright(c)2004-2008  By University of Cincinnati.
c            This code is intended for educational use only. No part of
c            the code can be used for any commercial applications/
c            distributions without prior written permission of the
c            original developer.
c
c----------------------------------------------------------------------
      implicit real*8(a-h,o-z)

      character*80 Prob_Title

      allocatable :: a(:,:),u(:),x(:,:),y(:,:),node(:,:),bc(:,:),
     &               dnorm(:,:),xfield(:,:),f(:),atu(:),itemp(:)
```

```
      open (5, file='input.dat',          status='old')
      open (6, file='output.dat',         status='unknown')
      open (7, file='phi_boundary.plt',   status='unknown')
      open (8, file='xy.plt',             status='unknown')
      open (9, file='phi_domain.plt',     status='unknown'

      call CPU_Time(time0)

c Read in initial data

      read(5,1) Prob_Title
      read(5,*) n, nfield
1     format(A80)

      write(6,1) Prob_Title
      write(*,1) Prob_Title
      write(6,*) ' Total number of elements =', n
      write(*,*) ' Total number of elements =', n
      write(6,*)
      write(*,*)

c Allocate the arrays

      allocate (a(n,n),u(n),x(2,n),y(2,n),node(2,n),bc(2,n),dnorm(2,n),
     &          xfield(2,nfield),f(nfield),atu(n),itemp(n))

c Input and prepare the BEM model

      call prep_model(n,x,y,bc,dnorm,node,xfield,nfield)

c Compute the right-hand-side vector b

      call bvector(u,x,y,bc,node,dnorm,n)

c Computer the coefficient matrix A

      call coefficient(a,n,x,y,bc,node,dnorm)

c Solve the system of equations Ax = b

c Use LAPACK direct solver (double precision, available at www.netlib.org)

          write(6,*) '   LAPACK direct solver is called ......'
```

```fortran
      write(*,*) '   LAPACK direct solver is called ......'
      call dgesv(n,1,a,n,itemp,u,n,info)
      write(6,*) '   LAPACK solver info =    ', info
      write(*,*) '   LAPACK solver info =    ', info

c Output the boundary solution

      write(6,*)
      write(6,*) ' Boundary Solution:'
      do i=1,n
         write(6,*) i, u(i)
         write(7,*) i, u(i)
      enddo

c Evaluate the potential field inside the domain and output the results

      call domain_field(nfield,xfield,f,n,x,y,bc,node,dnorm,u)

c Estimate the total CPU time

      call CPU_Time(time)
      write(*,*)
      write(*,*) ' Total CPU time used =', time-time0, '(sec)'
      write(6,*)
      write(6,*) ' Total CPU time used =', time-time0, '(sec)'

      stop
      end
c-------------------------------------------------------------------------
c Definition of the Variables:
c
c n               = total number of (middle) nodes (elements)
c x(2,n)          = coordinates of the nodes
c y(2,n)          = coordinates of the end points defining the elements
c node(2,n)       = element connectivity
c bc(2,n)         = bc(1,i) contains BC type, bc(2,i) BC value, for
c                   element i
c dnorm(2,n)      = normal of the elements
c a(n,n)          = matrix A
c u(n)            = first stores b vector; then solution vector of Ax = b
c nfield          = total number of the field points inside the domain
c xfield(2,nfield) = coordinates of the field points inside the domain
```

```
c f(nfield)        = values of potential at field points inside the domain
c atu(n)           = temp array for the solver
c itemp(n)         = temp array for the solver
c
c----------------------------------------------------------------------

      subroutine prep_model(n,x,y,bc,dnorm,node,xfield,nfield)
      implicit real*8(a-h,o-z)
      dimension x(2,*),y(2,*),bc(2,*),xfield(2,*),dnorm(2,*),node(2,*)

c Input the mesh data

      read(5,*)
      do i=1,n
         read(5,*) itemp, y(1,i), y(2,i)
      enddo

      read(5,*)
      do i=1,n
         read(5,*) itemp, node(1,i), node(2,i), bc(1,i), bc(2,i)
      enddo

c Input the field points inside the domain

      if (nfield .gt. 0) then
         read(5,*)
         do i=1,nfield
           read(5,*) itemp, xfield(1,i), xfield(2,i)
         enddo
      endif
c Compute mid-nodes and normals of the elements

      do i=1,n
         x(1,i) = (y(1,node(1,i)) + y(1,node(2,i)))*0.5d0
         x(2,i) = (y(2,node(1,i)) + y(2,node(2,i)))*0.5d0
         h1 =  y(2,node(2,i)) - y(2,node(1,i))
         h2 = -y(1,node(2,i)) + y(1,node(1,i))
         el = sqrt(h1**2 + h2**2)
         dnorm(1,i) = h1/el
         dnorm(2,i) = h2/el
      enddo
```

```
c Output nodal coordinates for plotting/checking

      do i = 1,n
         write(8,*) x(1,i), x(2,i)
      enddo
      return
      end


c------------------------------------------------------------------------


      subroutine bvector(u,x,y,bc,node,dnorm,n)

      implicit real*8(a-h,o-z)

      dimension u(*),x(2,*),y(2,*),bc(2,*),node(2,*),dnorm(2,*)

      data pi/3.141592653589793D0/

      pi2 = pi*2

      do i=1,n

         u(i) = 0.d0

      enddo

      do j=1,n            ! Loop on field points (Column)

         al   = sqrt((y(1,node(2,j)) - y(1,node(1,j)))**2 +
                     (y(2,node(2,j)) - y(2,node(1,j)))**2)  !    Element length

      do i=1,n            ! Loop on source points (Row)

c Compute parameters used in the formulas for the two intergals

            x11 = y(1,node(1,j)) - x(1,i)
            x21 = y(2,node(1,j)) - x(2,i)
            x12 = y(1,node(2,j)) - x(1,i)
            x22 = y(2,node(2,j)) - x(2,i)

            r1 =  sqrt(x11**2 + x21**2)
            r2 =  sqrt(x12**2 + x22**2)
            d  =  x11*dnorm(1,j) + x21*dnorm(2,j)
```

```fortran
      t1 = -x11*dnorm(2,j) + x21*dnorm(1,j)
      t2 = -x12*dnorm(2,j) + x22*dnorm(1,j)

      ds = abs(d)
      theta1 = datan2(t1,ds)
      theta2 = datan2(t2,ds)
      dtheta = theta2 - theta1

      aa = (-dtheta*ds + al + t1*log(r1) - t2*log(r2))/pi2
      if(d .lt. 0.d0) dtheta = -dtheta
      bb = -dtheta/pi2
      if(i .eq. j) bb = 0.5

      if(bc(1,j).eq.1.) u(i) = u(i) - bb*bc(2,j)      ! Potential is given
      if(bc(1,j).eq.2.) u(i) = u(i) + aa*bc(2,j)      ! Derivative is given

   enddo
   enddo
   return
   end
c-------------------------------------------------------------------------

   subroutine coefficient(a,n,x,y,bc,node,dnorm)

   implicit real*8(a-h,o-z)

   dimension a(n,n),x(2,*),y(2,*),bc(2,*),node(2,*),dnorm(2,*)

   data pi/3.141592653589793D0/

    pi2 = pi*2

   do j=1,n
   do i=1,n
      a(i,j) = 0.d0
   enddo
   enddo

   do j=1,n           ! Loop on field points (Column)

  al   = sqrt((y(1,node(2,j)) - y(1,node(1,j)))**2 +
           (y(2,node(2,j)) - y(2,node(1,j)))**2)      ! Element length
```

```
      do i=1,n            ! Loop on source points (Row)


         x11 = y(1,node(1,j)) - x(1,i)
         x21 = y(2,node(1,j)) - x(2,i)
         x12 = y(1,node(2,j)) - x(1,i)
         x22 = y(2,node(2,j)) - x(2,i)


         r1 =  sqrt(x11**2 + x21**2)
         r2 =  sqrt(x12**2 + x22**2)
         d  =  x11*dnorm(1,j) + x21*dnorm(2,j)
         t1 = -x11*dnorm(2,j) + x21*dnorm(1,j)
         t2 = -x12*dnorm(2,j) + x22*dnorm(1,j)


         ds = abs(d)
         theta1 = datan2(t1,ds)
         theta2 = datan2(t2,ds)
         dtheta = theta2 - theta1


         aa = (-dtheta*ds + al + t1*log(r1) - t2*log(r2))/pi2
         if(d .lt. 0.d0) dtheta = -dtheta
         bb = -dtheta/pi2

      if(i.ne.j)    then
         if(bc(1,j).eq.1.) a(i,j) = a(i,j) - aa
         if(bc(1,j).eq.2.) a(i,j) = a(i,j) + bb
         endif

      if(i.eq.j)    then
         if(bc(1,j).eq.1.) a(i,j) = a(i,j) - aa
         if(bc(1,j).eq.2.) a(i,j) = a(i,j) + 0.5d0
         endif

      enddo
      enddo

      return
      end


c---------------------------------------------------------------------


      subroutine domain_field(nfield,xfield,f,n,x,y,bc,node,dnorm,u)

      implicit real*8(a-h,o-z)
```

```fortran
      dimension xfield(2,*), f(*), x(2,*),y(2,*),bc(2,*),node(2,*),
     &          dnorm(2,*),u(*)

      data      pi/3.141592653589793D0/

      pi2 = pi*2.d0

      do i=1,nfield
         f(i) = 0.d0
      enddo

      do j=1,n        ! Loop over all elements

      if(bc(1,j).eq.1) then
           f0  = bc(2,j)
           df0 = u(j)
      else if(bc(1,j).eq.2) then
           f0  = u(j)
           df0 = bc(2,j)
      endif

        al   = sqrt((y(1,node(2,j)) - y(1,node(1,j)))**2 +
     &              (y(2,node(2,j)) - y(2,node(1,j)))**2)    ! Element length

       do i=1,nfield     ! Loop over all field points inside the domain

         x11 = y(1,node(1,j)) - xfield(1,i)
         x21 = y(2,node(1,j)) - xfield(2,i)
         x12 = y(1,node(2,j)) - xfield(1,i)
         x22 = y(2,node(2,j)) - xfield(2,i)

         r1 =  sqrt(x11**2 + x21**2)
         r2 =  sqrt(x12**2 + x22**2)
         d  =  x11*dnorm(1,j) + x21*dnorm(2,j)
         t1 = -x11*dnorm(2,j) + x21*dnorm(1,j)
         t2 = -x12*dnorm(2,j) + x22*dnorm(1,j)

         ds = abs(d)
         theta1 = datan2(t1,ds)
         theta2 = datan2(t2,ds)
         dtheta = theta2 - theta1

         aa = (-dtheta*ds + al + t1*log(r1) - t2*log(r2))/pi2
```

```
          if(d .lt. 0.d0) dtheta = −dtheta
          bb = −dtheta/pi2


          f(i) = f(i) + aa*df0 − bb*f0


      enddo
      enddo


c Output results


      do i=1,nfield
        write(9,20) xfield(1,i), f(i)
      enddo
20    format(1x, 4E18.8)


      return
      end
c--------------------------------------------------------------------
```

B.2 A Fortran Code of the Fast Multipole Boundary Element Method for 2D Potential Problems

The following is a list of the source code written in Fortran for the program discussed in Section 3.3 for 2D potential problems using the fast multipole BEM. The GMRES iterative solver *dgmres* from the SLATEC package can be downloaded from the website www.netlib.org.

```
c--------------------------------------------------------------------


c Program:    2D_Potential_FMM - A fast multipole boundary element
c             Method (BEM) code for analyzing large-scale, general 2D
c             potential problems (governed by Laplace equation)
c             using constant elements.
c
c Developers: Dr. Naoshi Nishimura at Kyoto University, Japan;
c             Dr. Yijun Liu at the University of Cincinnati, Cincinnati, OH.
c
c Version:    V.1.20.
c Released:   October 1, 2008.
c
c Copyright(c)2004-2008 By Kyoto University and University of Cincinnati.
```

```
c              This code is intended for educational use only. No part of
c              the code can be used for any commercial applications/
c              distributions without prior written permissions of the
c              original developers.
c
c--------------------------------------------------------------------------

       implicit real*8(a-h,o-z)

       integer, allocatable    :: ia(:)
       complex*16, allocatable :: am(:)

       character*80 Prob_Title

       call CPU_Time(time0)

       open (4, file='input.fmm',         status='old')
       open (5, file='input.dat',         status='old')
       open (3, file='output.dat',        status='unknown')
       open (7, file='phi_boundary.plt',  status='unknown')
       open (8, file='xy.plt',            status='unknown')
       open (9, file='phi_domain.plt',    status='unknown')

c Input the parameters

       read(4,*)        maxl, levmx, nexp, ntylr, tolerance
       read(4,*)        maxia, ncellmx, nleafmx, mxl, nwksz
       read(5,'(a80)')  Prob_Title
       read(5,*)        n, nfield
       write(3,'(a80)') Prob_Title
       write(*,'(a80)') Prob_Title

c Estimate the maximum numbers of the cells and leaves,
c and size of the preconditioning matrix, etc.

       if(ncellmx.le.0) ncellmx = max(4*n/maxl,100)
       if(nleafmx.le.0) nleafmx = max(ncellmx/2,100)
       if(nwksz.le.0)   nwksz   = maxl*maxl*nleafmx
       ligw  = 20
       lrgw  = 1+n*(mxl+6)+mxl*(mxl+3)
       iwksz = n+3*nleafmx+1

       allocate (ia(maxia))
```

```
c Load the addresses (pointers) associated with the locations of the
c variables to be stored in the large array "am"

      call lpointer(lp, ln, maxia, ia, n, nexp, ntylr, ncellmx,
     &        levmx, ligw, lrgw, nwksz, iwksz, nfield,
     &        l_n,        l_x,        l_y,        l_node, l_dnorm,
     &        l_bc,       l_a,        l_b,        l_xmax,
     &        l_xmin,     l_ymax,     l_ymin,     l_ielem, l_itree,
     &        l_level,    l_loct,     l_numt,     l_ifath, l_lowlev,
     &        l_maxl,     l_levmx,    l_nexp,     l_ntylr, l_tolerance,
     &        l_ncellmx,  l_nleafmx,  l_mxl,      l_u,     l_ax,
     &        l_sb,       l_sx,       l_ligw,     l_lrgw,  l_igwk,
     &        l_rgwk,     l_nwksz,    l_iwksz,    l_rwork, l_iwork,
     &        l_xfield,   l_nfield,   l_f)


c Estimate the memory usage

      maxa = lp
      write(3,100) maxa*16/1.D6
      write(*,100) maxa*16/1.D6
100   format(' Memory size of the large block am  =', f12.1,' Mb'/)


c Allocate the large block 'am'

      allocate (am(maxa))


c Assign the parameters to the array am()

      call assigni(n,         am(l_n      ))
      call assigni(maxl,      am(l_maxl   ))
      call assigni(levmx,     am(l_levmx  ))
      call assigni(nexp,      am(l_nexp   ))
      call assigni(ntylr,     am(l_ntylr  ))
      call assignd(tolerance, am(l_tolerance))
      call assigni(ncellmx,   am(l_ncellmx))
      call assigni(nleafmx,   am(l_nleafmx))
      call assigni(mxl,       am(l_mxl    ))
      call assigni(ligw,      am(l_ligw   ))
      call assigni(lrgw,      am(l_lrgw   ))
      call assigni(nwksz,     am(l_nwksz  ))
      call assigni(iwksz,     am(l_iwksz  ))
      call assigni(nfield,    am(l_nfield))
```

```
c Call the FMM BEM main program

      call fmmmain(maxa,    maxia,        am,          ia,
     &     am(l_n),       am(l_x),       am(l_y),      am(l_node),
     &     am(l_dnorm),   am(l_bc),      am(l_a),      am(l_b),
     &     am(l_xmax),    am(l_xmin),    am(l_ymax),   am(l_ymin),
     &     am(l_ielem),   am(l_itree),   am(l_level),  am(l_loct),
     &     am(l_numt),    am(l_ifath),   am(l_lowlev), am(l_maxl),
     &     am(l_levmx),   am(l_nexp),    am(l_ntylr),  am(l_tolerance),
     &     am(l_ncellmx), am(l_nleafmx), am(l_mxl),    am(l_u),
     &     am(l_ax),      am(l_nfield),  am(l_xfield), am(l_f),
     &     am(l_sb),      am(l_sx),      am(l_igwk),   am(l_rgwk),
     &     am(l_ligw),    am(l_lrgw),    am(l_nwksz),  am(l_iwksz),
     &     am(l_rwork),   am(l_iwork))

c Estimate the total CPU time

      call CPU_Time(time)
      write(3,*)
      write(*,*)
      write(3,*) ' Total CPU time used =', time-time0, '(sec)'
      write(*,*) ' Total CPU time used =', time-time0, '(sec)'

      stop
      end


c---------------------------------------------------------------------------
c Definition of Variables:
c
c maxa     = maximum size of the array am
c maxia    = maximum number of variables allowed
c am       = a large array storing the variables for the SLATEC GMRES solver
c ia       = an array storing the locations of the variables in the array am
c
c n        = number of elements (= number of nodes)
c x        = coordinates of the nodes
c y        = coordinates of the end points of the elements
c node     = element connectivity
c dnorm    = normal at each node
c bc       = BC type and value
c
c a        = multipole expansion moments
```

```
c b           = local expansion coefficients
c xmax,xmin = maximum and minimum x coordinate
c ymax,ymin = maximum and minimum y coordinate
c ielem      = ielem(i) gives the original element number for i-th element in
c               the quad-tree structure
c itree      = itree(c) gives the cell location of c-th cell within each
c               tree level
c loct       = elements included in the c-th cell are listed starting at
c               the loct(c)-th place in the array ielem
c numt       = numt(c) gives the number of elements included in the c-th cell
c ifath      = ifath(c) gives the number of the parent cell of the c-th cell
c level      = level l cells start at the level(l)-th cell in the tree
c lowlev     = number of the tree levels
c
c maxl       = maximum number of elements allowed in a leaf
c levmx      = maximum number of levels allowed in the tree structure
c nexp       = number of terms in multipole expansion
c ntylr      = number of terms in local expansion
c tolerance = GMRES solution convergence tolerance
c ncellmx    = maximum number of cells allowed in the tree
c nleafmx    = maximum number of leaves allowed in the tree
c mxl        = maximum dimension of Krylov subspace (used in GMRES)
c
c u          = first stores b vector; then solution vector of system Ax = b
c ax         = resulting vector of multiplication Ax
c nfield     = number of the field points inside the domain
c xfield     = coordinates of the field points inside the domain
c f          = values of the potential at the field points inside the domain
c
c The following variables and arrays are used in the SLATEC GMRES solver:
c    sb,sx,igwk,rgwk,ligw,lrgw,nwksz,iwksz,rwork,iwork
c
c-------------------------------------------------------------------------

      subroutine lpointer(lp, ln, maxia, ia, n, nexp, ntylr, ncellmx,
     &              levmx, ligw, lrgw, nwksz, iwksz, nfield,
     &              l_n,      l_x,      l_y,      l_node,  l_dnorm,
     &              l_bc,     l_a,      l_b,      l_xmax,
     &              l_xmin,   l_ymax,   l_ymin,   l_ielem, l_itree,
     &              l_level,  l_loct,   l_numt,   l_ifath, l_lowlev,
     &              l_maxl,   l_levmx,  l_nexp,   l_ntylr, l_tolerance,
     &              l_ncellmx, l_nleafmx, l_mxl,   l_u,     l_ax,
     &              l_sb,     l_sx,     l_ligw,   l_lrgw,  l_igwk,
```

```
&                   l_rgwk,    l_nwksz,   l_iwksz, l_rwork, l_iwork,
&                   l_xfield,  l_nfield,  l_f)

dimension ia(maxia)

lp  = 1

l_n          = l_address( 1,maxia,ia,lp,4,1)
l_x          = l_address( 2,maxia,ia,lp,8,n*2)
l_y          = l_address( 3,maxia,ia,lp,8,n*2)
l_node       = l_address( 4,maxia,ia,lp,4,n*2)
l_dnorm      = l_address( 5,maxia,ia,lp,8,n*2)
l_bc         = l_address( 6,maxia,ia,lp,8,n*2)
l_a          = l_address( 7,maxia,ia,lp,16,(nexp+1)*ncellmx)
l_b          = l_address( 8,maxia,ia,lp,16,(ntylr+1)*ncellmx)
l_xmax       = l_address( 9,maxia,ia,lp,8,1)
l_xmin       = l_address(10,maxia,ia,lp,8,1)
l_ymax       = l_address(11,maxia,ia,lp,8,1)
l_ymin       = l_address(12,maxia,ia,lp,8,1)
l_ielem      = l_address(13,maxia,ia,lp,4,n)
l_itree      = l_address(14,maxia,ia,lp,4,ncellmx)
l_level      = l_address(15,maxia,ia,lp,4,levmx+1)
l_loct       = l_address(16,maxia,ia,lp,4,ncellmx)
l_numt       = l_address(17,maxia,ia,lp,4,ncellmx)
l_ifath      = l_address(18,maxia,ia,lp,4,ncellmx)
l_lowlev     = l_address(19,maxia,ia,lp,4,1)
l_maxl       = l_address(20,maxia,ia,lp,4,1)
l_levmx      = l_address(21,maxia,ia,lp,4,1)
l_nexp       = l_address(22,maxia,ia,lp,4,1)
l_ntylr      = l_address(23,maxia,ia,lp,4,1)
l_tolerance  = l_address(24,maxia,ia,lp,8,1)
l_ncellmx    = l_address(25,maxia,ia,lp,4,1)
l_nleafmx    = l_address(26,maxia,ia,lp,4,1)
l_mxl        = l_address(27,maxia,ia,lp,4,1)
l_u          = l_address(28,maxia,ia,lp,8,n)
l_ax         = l_address(29,maxia,ia,lp,8,n)
l_sb         = l_address(30,maxia,ia,lp,8,n)
l_sx         = l_address(31,maxia,ia,lp,8,n)
l_ligw       = l_address(32,maxia,ia,lp,4,1)
l_lrgw       = l_address(33,maxia,ia,lp,4,1)
l_igwk       = l_address(34,maxia,ia,lp,4,ligw)
l_rgwk       = l_address(35,maxia,ia,lp,8,lrgw)
```

```fortran
      l_nwksz    = l_address(36,maxia,ia,lp,4,1)
      l_iwksz    = l_address(37,maxia,ia,lp,4,1)
      l_rwork    = l_address(38,maxia,ia,lp,8,nwksz)
      l_iwork    = l_address(39,maxia,ia,lp,4,iwksz)
      l_xfield   = l_address(40,maxia,ia,lp,8,nfield*2)
      l_nfield   = l_address(41,maxia,ia,lp,4,1)
      l_f        = l_address(42,maxia,ia,lp,8,nfield)

      return
      end


c-------------------------------------------------------------------------


      integer function l_address(ln,maxia,ia,lp,ibyte,length)

      dimension ia(maxia)

      l_address = lp
      ia(ln)    = lp

      iu  = 16
      inc = (ibyte*length-1)/iu+1
      lp  = lp+inc
      if(ln .gt. maxia) then
        write(*,*)'!Specified # of variables maxia',maxia,'is too small'
        stop
      endif

      return
      end


c-------------------------------------------------------------------------


      subroutine assigni(i,ii)

      integer i,ii
      ii = i

      return
      end

      subroutine assignd(d,dd)
```

```
      real*8 d,dd
      dd = d

      return
      end
```

c--

```
      subroutine fmmmain(maxa, maxia, am, ia, n,x,y,node,dnorm,bc,
     &           a,b, xmax,xmin,ymax,ymin,ielem,itree,level,loct,numt,
     &           ifath,lowlev,maxl,levmx,nexp,ntylr,tolerance,ncellmx,
     &           nleafmx,mxl,u,ax,nfield,xfield,f,sb,sx,igwk,rgwk,
     &           ligw,lrgw,nwksz,iwksz,rwork,iwork)

      implicit real*8(a-h,o-z)
      complex*16 am(maxa), a,b
      dimension  ia(maxia),ja(1),a(0:nexp,ncellmx),b(0:ntylr,ncellmx),
     &           x(2,n),y(2,n),node(2,n),dnorm(2,n),bc(2,n),
     &           ielem(n),itree(ncellmx),level(0:levmx),loct(ncellmx),
     &           numt(ncellmx),ifath(ncellmx), u(n),ax(n),sb(n),sx(n),
     &           igwk(ligw),rgwk(lrgw),rwork(nwksz),iwork(iwksz),
     &           xfield(2,nfield),f(nfield)

      external  matvec, msolve

c Input parameters and prepare the BEM model

      call prep_model(n,x,y,node,bc,dnorm,xfield,nfield,maxl,levmx,
     &                nexp,ntylr,tolerance,xmin,xmax,ymin,ymax)
c Generate the quad-tree structure for the elements

      call tree(n,x,xmax,xmin,ymax,ymin,ielem,itree,level,loct,numt,
     &          ifath,lowlev,maxl,levmx,ncellmx,nleafmx,nwksz,iwork)

c Compute the right-hand-side vector b with the FMM

      call fmmbvector(n,x,y,node,dnorm,bc,u,ax,a,b,xmax,xmin,ymax,ymin,
     &                ielem,itree,level,loct,numt,ifath,
     &                nexp,ntylr,ncellmx,lowlev,maxl,rwork,iwork)

c Solve the BEM system of equations Ax=b with the fast multipole BEM

c Prepare parameters for calling the iterative solver GMRES
c (SLATEC GMRES solver is used, which is available at www.netlib.org.
```

```
c See the documentation for the SLATEC GMRES solver for more information
c about the following related parameters)

      nelt    = 1
      isym    = 0
      itol    = 0
      tol     = tolerance
      iunit   = 3
      igwk(1) = mxl
      igwk(2) = mxl
      igwk(3) = 0
      igwk(4) = 1
      igwk(5) = 10
      do i=1,n
        ax(i) = 0.d0
      enddo

      write(*,*) 'Call Equation Solver GMRES ...'
      call dgmres(n,u,ax, nelt,ia,ja,am,isym, matvec,msolve,itol,tol,
     &            itmax,iter,er,ierr,iunit,sb,sx,rgwk,lrgw,igwk,ligw,
     &            rwork,iwork)
      write(3,*) ' Error indicator from GMRES:', ierr
      write(*,*) ' Error indicator from GMRES:', ierr

c Output the boundary solution

      do i=1,n
        u(ielem(i)) = ax(i)
      enddo

      write(3,*) ' Fast Multipole BEM Solution:'
      do i=1,n
        write(3,*) i, u(i)
        write(7,*) i, u(i)
      enddo

c Evaluate the field inside the domain and output the results

      call domain_field(nfield,xfield,f,n,x,y,bc,node,dnorm,u)

      return
      end
c------------------------------------------------------------------------
```

```
      subroutine prep_model(n,x,y,node,bc,dnorm,xfield,nfield,maxl,

     &                      levmx,nexp,ntylr,tolerance,
     &                      xmin,xmax,ymin,ymax)

      implicit real*8(a-h,o-z)

      dimension x(2,*),y(2,*),node(2,*),bc(2,*),dnorm(2,*),xfield(2,*)

      write(*,2) n, maxl, levmx, nexp, tolerance
      write(3,2) n, maxl, levmx, nexp, tolerance
2     format(' Total number of elements             =', I12
     &      /' Max. number of elements in a leaf =', I12
     &      /' Max. number of tree levels          =', I12
     &      /' Number of terms used in expansions =', I12
     &      /' Tolerance for convergence           =', D12.3)
      write(*,*)
      write(3,*)

c Input the mesh data

      read(5,*)
      do i=1,n
        read(5,*) itemp, y(1,i), y(2,i)
      enddo

      read(5,*)
      do i=1,n
        read(5,*) itemp, node(1,i), node(2,i), bc(1,i), bc(2,i)
      enddo

c Input the field points inside the domain

      if (nfield .gt. 0)  then
      read(5,*)
      do i=1,nfield
        read(5,*) itemp, xfield(1,i), xfield(2,i)
      enddo
      endif

c Compute mid-nodes and normals of the elements
```

```fortran
      do i=1,n
        x(1,i) = (y(1,node(1,i)) + y(1,node(2,i)))*0.5
        x(2,i) = (y(2,node(1,i)) + y(2,node(2,i)))*0.5
        h1 =  y(2,node(2,i)) - y(2,node(1,i))
        h2 = -y(1,node(2,i)) + y(1,node(1,i))
        el = sqrt(h1**2 + h2**2)
        dnorm(1,i) = h1/el
        dnorm(2,i) = h2/el
      enddo

c Determine the square bounding the problem domain (Largest cell used in FMM)

      xmin=x(1,1)
      xmax=x(1,1)
      ymin=x(2,1)
      ymax=x(2,1)

      do 10 i=2,n
         if(x(1,i).le.xmin) then
            xmin=x(1,i)
         elseif(x(1,i).ge.xmax) then
            xmax=x(1,i)
         endif
         if(x(2,i).le.ymin) then
            ymin=x(2,i)
         elseif(x(2,i).ge.ymax) then
            ymax=x(2,i)
         endif
10    continue

      scale = 1.05d0    ! Make the square slightly larger
      xyd   = max(xmax-xmin,ymax-ymin)/2.d0
      xyd   = xyd*scale
      cx    = (xmin+xmax)/2.d0
      cy    = (ymin+ymax)/2.d0
      xmin  = cx-xyd
      xmax  = cx+xyd
      ymin  = cy-xyd
      ymax  = cy+xyd

c Output nodal coordinates for plotting

      do i = 1,n
```

```
      write(8,*) x(1,i), x(2,i)
    enddo
    return
    end


c-------------------------------------------------------------------------


    subroutine tree(n,x,xmax,xmin,ymax,ymin,ielem,itree,level,loct,
   &               numt,ifath,lowlev,maxl,levmx,ncellmx,nleafmx,
   &               nwksz,iwork)

    implicit real*8(a-h,o-z)
    complex*16 a,b

    dimension x(2,*),ielem(*),itree(*),level(0:*),loct(*),numt(*),
   &          ifath(*), iwork(*), nwk(4)

    do i=1,n
      ielem(i) = i          ! Store the original element numbers in ielem
    enddo

c For the level 0 cell (largest cell)

    itree(1) = 0
    level(0) = 1
    level(1) = 2
    loct(1)  = 1
    ifath(1) = 1
    numt(1)  = n
    ndivx    = 1
    lowlev   = 1
    nleaf    = 0
    nswa     = 0

c For cells on level 1 to the lowest level (leaves)

    do 10 lev=1,levmx
      levp        = lev-1
      levn        = lev+1
      level(levn) = level(lev)
      if(level(lev).eq.level(levp)) goto 900
      ndivxp = ndivx
      ndivx  = 2*ndivxp
```

```
        dxp    = (xmax−xmin)/ndivxp      ! Parent cell size
        dyp    = (ymax−ymin)/ndivxp


        do 11 inp=level(levp),level(lev)−1
           itrp = itree(inp)
           if(numt(inp).gt.maxl .or.
     &         (lev.le.2 .and. numt(inp).ne.0) ) then
              itrpx = mod(itrp,ndivxp)
              itrpy = itrp/ndivxp
              xsep  = xmin+(itrpx + 0.5d0)*dxp
              ysep  = ymin+(itrpy + 0.5d0)*dyp

              call bisec(x,ielem(loct(inp)),numt(inp),ysep,nsepy, 2)
              call bisec(x,ielem(loct(inp)),nsepy−1,   xsep,nsepx1,1)
              call bisec(x,ielem(loct(inp)+nsepy−1),
     &                   numt(inp)-nsepy+1,xsep,nsepx2,1)
              nwk(1) = nsepx1−1
              nwk(2) = nsepy−nsepx1
              nwk(3) = nsepx2−1
              nwk(4) = numt(inp)−nsepy−nsepx2+2
              locc   = loct(inp)
              do 12 icldy=0,1
              do 12 icldx=0,1
                  icld = icldy*2+icldx+1

                 if(nwk(icld).gt.0) then
                    nrel = level(levn)
                    if(nrel.gt.ncellmx) then
                      write(*,*) " ncellmx error"
                      stop
                    endif
                    itree(nrel) = ((itrpy*2+icldy)*ndivxp + itrpx)*2
     &                            +icldx
                    loct(nrel)  = locc
                    numt(nrel)  = nwk(icld)
                    ifath(nrel) = inp
                    lowlev=lev

c Leaves:
                    if((lev.ne.1) .and.
     &                  (numt(nrel).le.maxl .or. lev.eq.levmx)) then
                       nleaf = nleaf+1
                         if(nleaf.gt.nleafmx)  then
```

```
                                  write(*,*) " nleafmx error"
                                stop
                              endif
                          nleaf3           = nleaf*3 - 1
                          iwork(nleaf3)    = nrel     ! Store cell
                                                         number (icell)
                          iwork(nleaf3+1) = nswa + 1  ! Location of
                                                          pre-cond'er
                          iwork(nleaf3+2) = 1        ! Initial value of switch isw
                          nswa             = nswa + numt(nrel)**2
                          if(nswa.gt.nwksz) then
                            write(*,*) " nwksz error"
                            stop
                          endif
                        endif


                        level(levn) = nrel + 1
                        locc        = locc + nwk(icld)
                      endif
12               continue
              endif
11       continue
10   continue

900  iwork(1) = nleaf                        ! Store number of leaves in iwork(1)

     write(3,*) ' Number of tree levels          =', lowlev
     write(*,*) ' Number of tree levels          =', lowlev
     write(3,*) ' Number of leaves               =', nleaf
     write(*,*) ' Number of leaves               =', nleaf
     write(3,*) ' Number of cells                =', nrel
     write(*,*) ' Number of cells                =', nrel
     write(3,*)
     write(*,*)

     return
     end


c----------------------------------------------------------------------

     subroutine bisec(x,ielem,n,xsep,nsep,ic)

     implicit        real*8(a-h,o-z)
```

```fortran
      dimension x(2,*),ielem(*)

      nsep = 1
      if(n.le.0) return

      do ifr=1,n
        if(x(ic,ielem(ifr)).le.xsep)  then
          if(ifr.ne.nsep) then
             istore      = ielem(nsep)
             ielem(nsep) = ielem(ifr)
             ielem(ifr)  = istore
          endif
          nsep = nsep + 1
        endif
      enddo

      return
      end


c-------------------------------------------------------------------------


      subroutine fmmbvector(n,x,y,node,dnorm,bc,u,ax,a,b,xmax,xmin,
     &               ymax,ymin,ielem,itree,level,loct,numt,ifath,
     &               nexp,ntylr,ncellmx,lowlev,maxl,rwork,iwork)

      implicit real*8(a-h,o-z)
      complex*16 a,b

      dimension a(0:nexp,ncellmx),b(0:ntylr,ncellmx),
     &          x(2,*),y(2,*),node(2,*),dnorm(2,*),bc(2,*),u(*),ax(*),
     &          ielem(*),itree(*),level(0:*),loct(*),numt(*),ifath(*),
     &          rwork(*),iwork(*)

c Switch the BC type

      do i=1,n
        if(bc(1,i) .eq. 1.) then
          bc(1,i) = 2.d0
        else
          bc(1,i) = 1.d0
        endif
      enddo
```

```
      do i=1,n
        u(i)  = bc(2,ielem(i))
        ax(i) = 0.d0
      enddo

c Apply the FMM to conpute the right-hand side vector b

      call upward(u,n,y,node,dnorm,bc,a,xmax,xmin,ymax,ymin,ielem,
     &          itree,level,loct,numt,ifath,nexp,ncellmx,lowlev,maxl)

      call dwnwrd(u,ax,n,x,y,node,dnorm,bc,a,b,xmax,xmin,ymax,ymin,
     &          ielem,itree,level,loct,numt,ifath,nexp,ntylr,ncellmx,
     &          lowlev,maxl,rwork,iwork)

c Store b vector in u and switch the BC type back

      do i=1,n
        u(i) = - ax(i)
        if(bc(1,i) .eq. 1.) then
          bc(1,i) = 2.d0
        else
          bc(1,i) = 1.d0
        endif
      enddo

      return
      end


c--------------------------------------------------------------------------


      subroutine matvec(n,u,ax,nelt,ia,ja,am,isym)

      implicit  real*8(a-h,o-z)
      complex*16 am, a, b

      dimension u(*),ax(*),ia(*),ja(*),am(*)

c Retrieve the pointers

      l_n       = ia(1)
      l_x       = ia(2)
      l_y       = ia(3)
      l_node    = ia(4)
```

```
l_dnorm     = ia(5)
l_bc        = ia(6)
l_a         = ia(7)
l_b         = ia(8)
l_xmax      = ia(9)
l_xmin      = ia(10)
l_ymax      = ia(11)
l_ymin      = ia(12)
l_ielem     = ia(13)
l_itree     = ia(14)
l_level     = ia(15)
l_loct      = ia(16)
l_numt      = ia(17)
l_ifath     = ia(18)
l_lowlev    = ia(19)
l_maxl      = ia(20)
l_levmx     = ia(21)
l_nexp      = ia(22)
l_ntylr     = ia(23)
l_tolerance = ia(24)
l_ncellmx   = ia(25)
l_nleafmx   = ia(26)
l_mxl       = ia(27)
l_u         = ia(28)
l_ax        = ia(29)
l_sb        = ia(30)
l_sx        = ia(31)
l_ligw      = ia(32)
l_lrgw      = ia(33)
l_igwk      = ia(34)
l_rgwk      = ia(35)
l_nwksz     = ia(36)
l_iwksz     = ia(37)
l_rwork     = ia(38)
l_iwork     = ia(39)

c Evaluate matrix-vector multiplication Ax using the fast multipole BEM

      call upward(u,        am(l_n),     am(l_y),      am(l_node),
     &      am(l_dnorm), am(l_bc),   am(l_a),      am(l_xmax),
     &      am(l_xmin),  am(l_ymax), am(l_ymin),   am(l_ielem),
     &      am(l_itree), am(l_level),am(l_loct),   am(l_numt),
     &      am(l_ifath), am(l_nexp), am(l_ncellmx),am(l_lowlev),
     &      am(l_maxl))
```

```
      call dwnwrd(u,ax,     am(l_n),        am(l_x),       am(l_y),
     &      am(l_node),  am(l_dnorm),  am(l_bc),    am(l_a),
     &      am(l_b),     am(l_xmax),   am(l_xmin),  am(l_ymax),
     &      am(l_ymin),  am(l_ielem),  am(l_itree), am(l_level),
     &      am(l_loct),  am(l_numt),   am(l_ifath), am(l_nexp),
     &      am(l_ntylr), am(l_ncellmx), am(l_lowlev), am(l_maxl),
     &      am(l_rwork), am(l_iwork))

      return
      end


c-------------------------------------------------------------------------

      subroutine msolve(n,r,z,nelt,ia,ja,am,isym,rwork,iwork)

      implicit real*8(a-h,o-z)
      complex*16 am(*)

      dimension  r(*),z(*),ia(*),ja(1),rwork(*),iwork(*)

c Load the pointers

      l_loct = ia(16)
      l_numt = ia(17)

c Compute the preconditioning matrix

      call msolveinv(r,z,rwork,iwork, am(l_loct), am(l_numt))

      return
      end

c-------------------------------------------------------------------------

      subroutine msolveinv(r,z,rwork,iwork,loct,numt)

      implicit real*8(a-h,o-z)

      dimension  r(*),z(*),iwork(*),rwork(*),loct(*),numt(*)

      nleaf = iwork(1)
      do l = 1,nleaf
        l3   = l*3-1
        inod = iwork(l3)
```

```fortran
      indr = iwork(13+1)
      indx = loct(inod)
      indi = indx+3*nleaf+1
      nr   = numt(inod)

  call dcopy(nr,r(indx),1,z(indx),1)
  call dluax(rwork(indr),nr,nr,z(indx),iwork(13+2),
 &           iwork(indi),icon)
  if(icon.ne.0) then
     write(*,*) " dluax error, icon =", icon
     stop
  endif
  iwork(13+2) = 2
  enddo

  return
  end


c-------------------------------------------------------------------------

c   This subroutine solves linear system of equations Ax=b by LU
decomposition.
c
c   a    .... given regular coefficient matrix.
c   k    .... given adjustable dimension for array a.
c   n    .... given order of matrix a.
c   b    .... given constant vector.
c   isw  .... given control information:
c              if 1, solve equations entirely.
c              if 2, solve equations with last LU-decomposed entries.
c   ip   .... auxiliary 1 dimensioned array, size is n.
c              transposition vector which represents
c              row-exchanging by partial pivoting.
c   icon.... resultant condition code.
c
c   Slave subroutines used (available at www.netlib.org):
c     dgetrf, dgetrs
c
c-------------------------------------------------------------------------

      subroutine dluax(a,k,n,b,isw,ip,icon)

      implicit real*8 (a-h,o-z)
      dimension a(k,n),b(n),ip(n)
      data ione/1/
```

```
      icon = 30000
      if(isw.eq.1) go to 1000
      if(isw.eq.2) go to 1100
      go to 8000
1000  call dgetrf(n,n,a,k,ip,icon)
      isw = 2
1100  call dgetrs('n',n,ione,a,k,ip,b,n,icon)
8000  continue

      return
      end


c-------------------------------------------------------------------------


      subroutine upward(u,n,y,node,dnorm,bc,a,xmax,xmin,ymax,ymin,ielem,
     &          itree,level,loct,numt,ifath,nexp,ncellmx,lowlev,maxl)

      implicit real*8(a-h,o-z)
      complex*16 a,b, z0,zi

      dimension a(0:nexp,ncellmx),
     &          y(2,*),node(2,*),dnorm(2,*),bc(2,*),u(*),
     &          ielem(*),itree(*),level(0:*),loct(*),numt(*),ifath(*)

      do i=1,level(lowlev+1)-1
      do k=0,nexp
        a(k,i) = (0.d0,0.d0)        ! Clear multipole moments
      enddo
      enddo

      do 10 lev=lowlev,2,-1         ! Loop from leaf to level 2 cells (Upward)

        ndivx = 2**lev
        dx    = (xmax-xmin)/ndivx   ! Determine cell size
        dy    = (ymax-ymin)/ndivx

        do 20 icell=level(lev),level(lev+1)-1  ! Loop for level l cells
          itr  = itree(icell)
          itrx = mod(itr,ndivx)
          itry = itr/ndivx                     ! Position of the cell
          cx   = xmin+(itrx + 0.5d0)*dx
          cy   = ymin+(itry + 0.5d0)*dy        ! Center of the cell
```

```fortran
c Multipole expansion

        if(numt(icell).le.maxl .or. lev.eq.lowlev) then  ! Compute moment
          call moment(a(0,icell),y,node,ielem(loct(icell)),
     &                 numt(icell),nexp,cx,cy,u(loct(icell)),
     &                              bc,dnorm)
        endif

c M2M translation

        if(lev.ne.2) then              ! Do M2M translation to form moments

        cxp = xmin+(int(itrx/2)*2 + 1)*dx
        cyp = ymin+(int(itry/2)*2 + 1)*dy    ! Center of parent cell
        z0  = cmplx(cx-cxp, cy-cyp)           ! (z_c - z_c')
        io  = ifath(icell)                    ! Cell no. of parent cell

        zi = (1.d0,0.d0)
        do k=0,nexp
          do m=k,nexp
            a(m,io) = a(m,io) + zi*a(m-k,icell)    ! Use M2M
          enddo
          zi = zi*z0/(k+1)
        enddo
      endif

20      continue

10      continue

      return
      end

c--------------------------------------------------------------------------

      subroutine moment(a,y,node,ielem,num,nexp,cx,cy,u,bc,dnorm)

      implicit real*8(a-h,o-y)
      implicit complex*16(z)

      complex*16 a(0:*)
      dimension  y(2,*),node(2,*),ielem(*),u(*), bc(2,*),dnorm(2,*)
```

```fortran
      do i=1,num     ! Over elements in the leaf

        nelm = ielem(i)                    ! Element number
        n1 = node(1,nelm)                  ! Two ends of the element
        n2 = node(2,nelm)
        z1 = cmplx(y(1,n1)-cx, y(2,n1)-cy)
        z2 = cmplx(y(1,n2)-cx, y(2,n2)-cy)
        zwbar = conjg(z2 - z1)
        zwbar = zwbar/abs(zwbar)           ! omega bar
        zp1   = z1*zwbar
        zp2   = z2*zwbar
        znorm = cmplx(dnorm(1,nelm),dnorm(2,nelm)) ! complex normal n
        if(bc(1,nelm) .eq. 1.d0) then      ! Assign values to phi and q
          phi = 0.D0
          q   = u(i)
        else if(bc(1,nelm) .eq. 2.d0) then
          phi = u(i)
          q   = 0.D0
        endif

c Compute moments:

      a(0) = a(0) - (zp2-zp1)*q            ! G kernel
      do k=1,nexp
        a(k) = a(k) + (zp2-zp1)*znorm*phi  ! F kernel
        zp1  = zp1*z1/(k+1)
        zp2  = zp2*z2/(k+1)
        a(k) = a(k) - (zp2-zp1)*q          ! G kernel
       enddo
      enddo

      return
      end

c------------------------------------------------------------------------

      subroutine dwnwrd(u,ax,n,x,y,node,dnorm,bc,a,b,xmax,xmin,
     &                  ymax,ymin,ielem,itree,level,loct,numt,ifath,
     &                  nexp,ntylr,ncellmx,lowlev,maxl,rwork,iwork)

      implicit real*8(a-h,o-z)
      complex*16 a,b, z0,zi,zo,zp
```

```
      dimension a(0:nexp,ncellmx),b(0:ntylr,ncellmx),x(2,*),y(2,*),
     &          node(2,*),dnorm(2,*),bc(2,*),u(*),ax(*),ielem(*),
     &          itree(*),level(0:*),loct(*),numt(*),ifath(*),
     &          rwork(*),iwork(*)

      data pi/3.141592653589793D0/

      pi2 = pi*2.d0

      do i=1,level(lowlev+1)-1
      do k=0,ntylr
         b(k,i) = (0.d0,0.d0)
      enddo
      enddo
      do i=1,n
         ax(i) = 0.d0
      enddo

      leaf = 0
      indr = 1
      indi = 1
      do 110 lev=2,lowlev        ! Downward from level 2 cells to leaf cells
         ndivx = 2**lev
         dx    = (xmax-xmin)/ndivx
         dy    = (ymax-ymin)/ndivx
         do 120 icell=level(lev),level(lev+1)-1 ! Loop for level l cells
            itr   = itree(icell)
            itrx  = mod(itr,ndivx)
            itry  = itr/ndivx                    ! Position of the cell
            cx    = xmin+(itrx + 0.5d0)*dx
            cy    = ymin+(itry + 0.5d0)*dy       ! Center of the cell
            itrxp = itrx/2
            itryp = itry/2
c From the parent cell (use L2L)

            if(lev.ne.2) then
               cxp = xmin+(itrxp*2+1)*dx
               cyp = ymin+(itryp*2+1)*dy         ! Center of the parent cell
               z0  = cmplx(cx-cxp, cy-cyp)
               io  = ifath(icell)                ! Cell no. of the parent cell
               zi  = (1.d0,0.d0)
               do k=0,ntylr
                  do m=0,ntylr-k
```

```
                   b(m,icell) = b(m,icell) + zi*b(k+m,io) ! L2L translation
                 enddo
                 zi = zi*z0/(k+1)
               enddo
             endif


       do 130 jcell=level(lev),level(lev+1)-1
         jtr   = itree(jcell)
         jtrx  = mod(jtr,ndivx)
         jtry  = jtr/ndivx
         jtrxp = jtrx/2
         jtryp = jtry/2


c The parents must be neighbours

         if(iabs(itrxp-jtrxp).gt.1 .or. iabs(itryp-jtryp).gt.1)
     &         goto 130


c For non-neighbours (cells in interaction list) (use M2L)

         if(iabs(itrx-jtrx).gt.1 .or. iabs(itry-jtry).gt.1) then
           ccx = xmin + (jtrx + 0.5d0)*dx
           ccy = ymin + (jtry + 0.5d0)*dy   ! Center of the j cell
z0 = cmplx(cx-ccx, cy-ccy)

           b(0,icell) = b(0,icell) - log(z0)*a(0,jcell)
           zo = 1.
           do m=1,nexp+ntylr
             zo   = zo/z0
             kmin = max(0,m-nexp)
             kmax = min(m,ntylr)
             sgn  = (-1.0)**kmin
             do k=kmin,kmax
               b(k,icell) = b(k,icell) + sgn*zo*a(m-k,jcell) ! M2L
               sgn = -sgn
             enddo
             zo = zo*m
           enddo


c Contribution from neighbouring leaves (use direct)

         elseif(numt(jcell).le.maxl .or.
     &          numt(icell).le.maxl .or. lev.eq.lowlev) then
           if(icell.eq.jcell) then
```

```
                          leaf  = leaf+1
                          leaf3 = leaf*3-1
                          if(iwork(leaf3).ne.icell) then
                              write(3,*)
      leaf,iwork(1),iwork(leaf3),icell,'check'
                              write(*,*) " icell error"
                              stop
                          endif
                          indr = iwork(leaf3+1)
                          indi = iwork(leaf3+2)
                      endif

                      call direct(ielem(loct(icell)),ielem(loct(jcell)),
    &                        node,x,y,numt(icell),numt(jcell),dnorm,
    &                        ax(loct(icell)),u(loct(jcell)),icell,jcell,
    &                        rwork(indr),indi,bc)        ! Direct integration
                  endif
130     continue

c Compute Ax if reach a leaf (Evaluate local expansion at each
c collocation point)

        if(numt(icell).le.maxl .or. lev.eq.lowlev) then
          fact = 1.d0
            do itylr=1,ntylr
              fact = fact/itylr
              b(itylr,icell) = b(itylr,icell)*fact
            enddo
          do in=1,numt(icell)
            inax = loct(icell) + in-1        ! Element number in the tree
            indx = ielem(inax)               ! Original element number
            zp   = b(ntylr,icell)
            z0   = cmplx(x(1,indx)-cx, x(2,indx)-cy)
            do itylr=ntylr-1,0,-1
              zp = zp*z0 + b(itylr,icell)    ! Local expansion
                enddo
            zp = zp/pi2
            ax(inax) = ax(inax) + dreal(zp) ! Array Ax
          enddo
        endif

120     continue
110     continue
```

```
      return
      end
c-------------------------------------------------------------------------

      subroutine direct(inod,jnod,node,x,y,ni,nj,dnorm,ax,u,icell,jcell,
     &                  amat,isw,bc)

      implicit real*8(a-h,o-z)

      dimension inod(*),jnod(*),node(2,*),x(2,*),y(2,*),
     &          dnorm(2,*),ax(*),u(*),amat(ni,*), bc(2,*)

      data      pi/3.141592653589793D0/

      pi2 = pi*2.d0

      do j = 1, nj

        jind = jnod(j)

        al = sqrt((y(1,node(1,jind))- y(1,node(2,jind)))**2 +
                  (y(2,node(1,jind)) - y(2,node(2,jind)))**2) ! Element length

      do i = 1, ni

        iind = inod(i)

        x11 = y(1,node(1,jind)) - x(1,iind)
        x21 = y(2,node(1,jind)) - x(2,iind)
        x12 = y(1,node(2,jind)) - x(1,iind)
        x22 = y(2,node(2,jind)) - x(2,iind)

        r1 =   sqrt(x11**2 + x21**2)
        r2 =   sqrt(x12**2 + x22**2)
        d  =   x11*dnorm(1,jind) + x21*dnorm(2,jind)
        t1 = -x11*dnorm(2,jind) + x21*dnorm(1,jind)
        t2 = -x12*dnorm(2,jind) + x22*dnorm(1,jind)

        ds = abs(d)
          dtheta = datan2(ds*al,ds**2+t1*t2)

        aa = (-dtheta*ds + al + t1*log(r1) - t2*log(r2))/pi2
        if(d .lt. 0.d0) dtheta = -dtheta
```

```fortran
      bb = -dtheta/pi2
      if(iind .eq. jind) bb = 0.5d0


   if(bc(1,jind) .eq. 1.) then
     ax(i) = ax(i) - aa*u(j)
     if(icell.eq.jcell .and. isw.eq.1) then     ! Store coefficients in
       amat(i,j) = - aa                         ! first iteration
     endif
   else if(bc(1,jind) .eq. 2.) then
     ax(i) = ax(i) + bb*u(j)
     if(icell.eq.jcell .and. isw.eq.1) then     ! Store coefficients in
       amat(i,j) =    bb                        ! first iteration
     endif
   endif

   enddo

  enddo

  return
  end


c------------------------------------------------------------------------------


      subroutine domain_field(nfield,xfield,f,n,x,y,bc,node,dnorm,u)

      implicit real*8(a-h,o-z)

      dimension xfield(2,*), f(*), x(2,*),y(2,*),bc(2,*),node(2,*),
     &          dnorm(2,*),u(*)

      data      pi/3.141592653589793D0/

      pi2 = pi*2.d0

      do i=1,nfield
        f(i) = 0.d0
      enddo

      do j=1,n        ! Loop over all elements

        if(bc(1,j).eq.1) then
          f0  = bc(2,j)
          df0 = u(j)
```

```
      else if(bc(1,j).eq.2) then
       f0  = u(j)
       df0 = bc(2,j)
      endif
    al    = sqrt((y(1,node(2,j)) - y(1,node(1,j)))**2 +
                  (y(2,node(2,j)) - y(2,node(1,j)))**2)      ! Element length

    do i=1,nfield     ! Loop over all field points inside the domain

     x11 = y(1,node(1,j)) - xfield(1,i)
     x21 = y(2,node(1,j)) - xfield(2,i)
     x12 = y(1,node(2,j)) - xfield(1,i)
     x22 = y(2,node(2,j)) - xfield(2,i)

     r1 =   sqrt(x11**2 + x21**2)
     r2 =   sqrt(x12**2 + x22**2)
     d  =   x11*dnorm(1,j) + x21*dnorm(2,j)
     t1 =  -x11*dnorm(2,j) + x21*dnorm(1,j)
     t2 =  -x12*dnorm(2,j) + x22*dnorm(1,j)

     ds = abs(d)
     theta1 = datan2(t1,ds)
     theta2 = datan2(t2,ds)
     dtheta = theta2 - theta1

     aa = (-dtheta*ds + al + t1*log(r1) - t2*log(r2))/pi2
     if(d .lt. 0.d0) dtheta = -dtheta
     bb = -dtheta/pi2

     f(i) = f(i) + aa*df0 - bb*f0

    enddo
    enddo

c Output results

    do i=1,nfield
      write(9,20) xfield(1,i), f(i)
    enddo
20    format(1x, 4E18.8)

    return
    end

c---------------------------------------------------------------------
```

B.3 Sample Input File and Parameter File

The following is a sample input file that can be used to run both the conventional BEM and the fast multipole BEM programs for 2D potential problems listed in the previous two sections. The model is for a square domain with dimensions of 1×1 and discretized with 20 constant line elements (five elements on each edge). A zero-potential BC is applied on the left edge and a potential of 100 is applied on the right edge. The upper and lower edges are applied with flux-free ($q = 0$) BCs. There are also 11 field points inside the domain where the potential will be evaluated after the solution on the boundary is obtained.

```
c-------------------------------------------------------------
c A Sample Input File (input.dat):
c----------------------------------------------------

A Square Plate with Linear Temperature                        30-MAR-04
          20        11                     ! No. of Elements, No. of Field Points
# Nodes (Node No., x-coordinate, y-coordinate):
1         0         0
2         0.2       0
3         0.4       0
4         0.6       0
5         0.8       0
6         1         0
7         1         0.2
8         1         0.4
9         1         0.6
10        1         0.8
11        1         1
12        0.8       1
13        0.6       1
14        0.4       1
15        0.2       1
16        0         1
17        0         0.8
18        0         0.6
19        0         0.4
20        0         0.2
# Elements and Boundary Conditions (Element No., Local Node 1, Local Node 2,
BC Type, BC Value):
1         1         2         2         0
2         2         3         2         0
```

```
3          3          4          2          0
4          4          5          2          0
5          5          6          2          0
6          6          7          1          100
7          7          8          1          100
8          8          9          1          100
9          9         10          1          100
10        10         11          1          100
11        11         12          2          0
12        12         13          2          0
13        13         14          2          0
14        14         15          2          0
15        15         16          2          0
16        16         17          1          0
17        17         18          1          0
18        18         19          1          0
19        19         20          1          0
20        20          1          1          0
# Field Points Inside Domain (Field Point No., x-coordinate, y-coordinate):
1          0.01       0.5
2          0.1        0.5
3          0.2        0.5
4          0.3        0.5
5          0.4        0.5
6          0.5        0.5
7          0.6        0.5
8          0.7        0.5
9          0.8        0.5
10         0.9        0.5
11         0.99       0.5
# End of File


c-----------------------------------------------------------
```

The following is the parameter file used only for the fast multipole BEM program for 2D potential problems. All the parameters are explained briefly in the file, which is used for the FMM and the iterative solver GMRES. In general, the values of these parameters in this file need not be changed when one is using the fast multipole BEM code, unless one wishes to use a different tolerance for convergence or number of terms in the multipole expansions.

```
c------------------------------------------------------------
c A Sample Parameter File (input.fmm, for 2D_Potential_FMM Program only):
c------------------------------------------------------------
    20    10    15    15    1.0E-8 ! maxl levmx  nexp   ntylr tolerance
    50 50000 50000    50  90000000 ! maxia ncellmx nleafmx mxl   nwksz

Definitions of the above parameters:

maxl:      maximum number of elements in a leaf
levmx:     maximum number of tree levels
nexp:      order of the fast multipole expansions (p)
ntylr:     order of the local expansions (= p, in general)
tolerance: tolerance for convergence used in the iterative solver
maxia:     maximum number of parameters
ncellmx:   maximum number of cells allowed in the tree
nleafmx:   maximum number of leaves allowed in the tree
mxl:       maximum dimension of Krylov subspace used in the iterative solver
nwksz:     size of the space used to store coefficients in preconditioner
           (use default in the code, if value = 0)
c------------------------------------------------------------
```

References

[1] M. A. Jaswon, "Integral equation methods in potential theory. I," *Proc. R. Soc. London A* **275**, 23–32 (1963).

[2] G. T. Symm, "Integral equation methods in potential theory. II," *Proc. R. Soc. London A* **275**, 33–46 (1963).

[3] M. A. Jaswon and A. R. Ponter, "An integral equation solution of the torsion problem," *Proc. R. Soc. London A* **273**, 237–246 (1963).

[4] F. J. Rizzo, "An integral equation approach to boundary value problems of classical elastostatics," *Q. Appl. Math.* **25**, 83–95 (1967).

[5] F. J. Rizzo and D. J. Shippy, "A formulation and solution procedure for the general non-homogeneous elastic inclusion problem," *Int. J. Solids Structures* **4**, 1161–1179 (1968).

[6] T. A. Cruse and F. J. Rizzo, "A direct formulation and numerical solution of the general transient elastodynamic problem – I," *J. Math. Anal. Appl.* **22**, 244–259 (1968).

[7] T. A. Cruse, "A direct formulation and numerical solution of the general transient elastodynamic problem – II," *J. Math. Anal. Appl.* **22**, 341–355 (1968).

[8] T. A. Cruse, "Numerical solutions in three dimensional elastostatics," *Int. J. Solids Structures* **5**, 1259–1274 (1969).

[9] F. J. Rizzo and D. J. Shippy, "A method for stress determination in plane anisotropic elastic bodies," *J. Composite Mater.* **4**, 36–61 (1970).

[10] F. J. Rizzo and D. J. Shippy, "A method of solution for certain problems of transient heat conduction," *AIAA J.* **8**, 2004–2009 (1970).

[11] F. J. Rizzo and D. J. Shippy, "An application of the correspondence principle of linear viscoelasticity theory," *SIAM J. Appl. Math.* **21**, 321–330 (1971).

[12] T. A. Cruse and W. V. Buren, "Three-dimensional elastic stress analysis of a fracture specimen with an edge crack," *Int. J. Fracture Mech.* **7**, 1–16 (1971).

[13] T. A. Cruse and J. L. Swedlow, "Formulation of boundary integral equations for three-dimensional elasto-plastic flow," *Int. J. Solids Structures* **7**, 1673–1683 (1971).

[14] T. A. Cruse, "Application of the boundary-integral equation method to three-dimensional stress analysis," *Computers Structures* **3**, 509–527 (1973).

[15] T. A. Cruse, "An improved boundary-integral equation method for three-dimensional elastic stress analysis," *Computers Structures* **4**, 741–754 (1974).

[16] T. A. Cruse and F. J. Rizzo, eds., *Boundary-Integral Equation Method: Computational Applications in Applied Mechanics* (AMD-ASME, New York 1975), Vol. 11.

[17] J. C. Lachat and J. O. Watson, "Effective numerical treatment of boundary integral equations: A formulation for three-dimensional elastostatics," *Int. J. Numer. Methods Eng.* **10**, 991–1005 (1976).

[18] F. J. Rizzo and D. J. Shippy, "An advanced boundary integral equation method for three-dimensional thermoelasticity," *Int. J. Numer. Methods Eng.* **11**, 1753–1768 (1977).

[19] M. Stippes and F. J. Rizzo, "A note on the body force integral of classical elastostatics," *Z. Angew. Math. Phys.* **28**, 339–341 (1977).

[20] R. B. Wilson and T. A. Cruse, "Efficient implementation of anisotropic three-dimensional boundary-integral equation stress analysis," *Int. J. Numer. Methods Eng.* **12**, 1383–1397 (1978).

[21] P. K. Banerjee and R. Butterfield, "Boundary element methods in geomechanics," in G. Gudehus, ed., *Finite Elements in Geomechanics* (Wiley, London, 1976), Chap. 16, pp. 529–570.

[22] P. K. Banerjee et al., eds., *Developments in Boundary Element Methods* (Elsevier Applied Science, London, 1979–1991), Vols. I–VII.

[23] C. A. Brebbia, *The Boundary Element Method for Engineers* (Pentech Press, London, 1978).

[24] P. K. Banerjee, *The Boundary Element Methods in Engineering*, 2nd ed. (McGraw-Hill, New York, 1994).

[25] S. Mukherjee, *Boundary Element Methods in Creep and Fracture* (Applied Science Publishers, New York, 1982).

[26] T. A. Cruse, *Boundary Element Analysis in Computational Fracture Mechanics* (Kluwer Academic, Dordrecht, The Netherlands, 1988).

[27] C. A. Brebbia and J. Dominguez, *Boundary Elements – An Introductory Course* (McGraw-Hill, New York, 1989).

[28] J. H. Kane, *Boundary Element Analysis in Engineering Continuum Mechanics* (Prentice-Hall, Englewood Cliffs, NJ, 1994).

[29] M. Bonnet, *Boundary Integral Equation Methods for Solids and Fluids* (Wiley, Chichester, UK, 1995).

[30] L. C. Wrobel, *The Boundary Element Method – Vol. 1, Applications in Thermo-Fluids and Acoustics* (Wiley, Chichester, UK, 2002).

[31] M. H. Aliabadi, *The Boundary Element Method – Vol. 2, Applications in Solids and Structures* (Wiley, Chichester, UK, 2002).

[32] S. Mukherjee and Y. X. Mukherjee, *Boundary Methods: Elements, Contours, and Nodes* (CRC, Boca Raton, FL, 2005).

[33] V. Rokhlin, "Rapid solution of integral equations of classical potential theory," *J. Comput. Phys.* **60**, 187–207 (1985).

[34] L. F. Greengard and V. Rokhlin, "A fast algorithm for particle simulations," *J. Comput. Phys.* **73**, 325–348 (1987).

[35] L. F. Greengard, *The Rapid Evaluation of Potential Fields in Particle Systems* (MIT Press, Cambridge, MA, 1988).

[36] A. P. Peirce and J. A. L. Napier, "A spectral multipole method for efficient solution of large-scale boundary element models in elastostatics," *Int. J. Numer. Methods Eng.* **38**, 4009–4034 (1995).

[37] J. E. Gomez and H. Power, "A multipole direct and indirect BEM for 2D cavity flow at low Reynolds number," *Eng. Anal. Boundary Elements* **19**, 17–31 (1997).

[38] Y. Fu, K. J. Klimkowski, G. J. Rodin, E. Berger, J. C. Browne, J. K. Singer, R. A. V. D. Geijn, and K. S. Vemaganti, "A fast solution method for three-dimensional many-particle problems of linear elasticity," *Int. J. Numer. Methods Eng.* **42**, 1215–1229 (1998).

[39] N. Nishimura, K. Yoshida, and S. Kobayashi, "A fast multipole boundary integral equation method for crack problems in 3D," *Eng. Anal. Boundary Elements* **23**, 97–105 (1999).

[40] A. A. Mammoli and M. S. Ingber, "Stokes flow around cylinders in a bounded two-dimensional domain using multipole-accelerated boundary element methods," *Int. J. Numer. Methods Eng.* **44**, 897–917 (1999).

[41] N. Nishimura, "Fast multipole accelerated boundary integral equation methods," *Appl. Mech. Rev.* **55**, 299–324 (2002).

[42] A. H. Zemanian, *Distribution Theory and Transform Analysis – An Introduction to Generalized Functions, with Applications* (Dover, New York, 1987).

[43] Y. C. Fung, *A First Course in Continuum Mechanics*, 3rd ed. (Prentice-Hall, Englewood Cliffs, NJ, 1994).

[44] J. Hadamard, *Lectures on Cauchy's Problem in Linear Partial Differential Equations* (Yale University Press, New Haven, CT, 1923).

[45] P. A. Martin and F. J. Rizzo, "Hypersingular integrals: How smooth must the density be?" *Int. J. Numer. Methods Eng.* **39**, 687–704 (1996).

[46] Y. J. Liu and T. J. Rudolphi, "Some identities for fundamental solutions and their applications to weakly-singular boundary element formulations," *Eng. Anal. Boundary Elements* **8**, 301–311 (1991).

[47] Y. J. Liu and T. J. Rudolphi, "New identities for fundamental solutions and their applications to non-singular boundary element formulations," *Comput. Mech.* **24**, 286–292 (1999).

[48] Y. J. Liu, "On the simple-solution method and non-singular nature of the BIE/BEM – A review and some new results," *Eng. Anal. Boundary Elements* **24**, 787–793 (2000).

[49] G. Krishnasamy, F. J. Rizzo, and Y. J. Liu, "Boundary integral equations for thin bodies," *Int. J. Numer. Methods Eng.* **37**, 107–121 (1994).

[50] Y. J. Liu and F. J. Rizzo, "A weakly-singular form of the hypersingular boundary integral equation applied to 3-D acoustic wave problems," *Comput. Methods Appl. Mech. Eng.* **96**, 271–287 (1992).

[51] Y. J. Liu and S. H. Chen, "A new form of the hypersingular boundary integral equation for 3-D acoustics and its implementation with C° boundary elements," *Comput. Methods Appl. Mech. Eng.* **173**, 3–4, 375–386 (1999).

[52] Y. J. Liu and F. J. Rizzo, "Hypersingular boundary integral equations for radiation and scattering of elastic waves in three dimensions," *Comput. Methods Appl. Mech. Eng.* **107**, 131–144 (1993).

[53] Y. J. Liu, D. M. Zhang, and F. J. Rizzo, "Nearly singular and hypersingular integrals in the boundary element method," in: C. A. Brebbia and J. J. Rencis, eds., *Boundary Elements XV* (Computational Mechanics Publications, Worcester, MA, 1993), pp. 453–468.

[54] X. L. Chen and Y. J. Liu, "An advanced 3-D boundary element method for characterizations of composite materials," *Eng. Anal. Boundary Elements* **29**, 513–523 (2005).

[55] P. W. Partridge, C. A. Brebbia, and L. C. Wrobel, *The Dual Reciprocity Boundary Element Method* (Computational Mechanics Publications, Southampton, UK, 1992).

[56] O. D. Kellogg, *Foundations of Potential Theory* (Dover, New York, 1953).

[57] Y. J. Liu, "Dual BIE approaches for modeling electrostatic MEMS problems with thin beams and accelerated by the fast multipole method," *Eng. Anal. Boundary Elements* **30**, 940–948 (2006).

[58] W. H. Hayt and J. A. Buck, *Engineering Electromagnetics* (McGraw-Hill, London, 2001).

[59] Y. J. Liu and L. Shen, "A dual BIE approach for large-scale modeling of 3-D electrostatic problems with the fast multipole boundary element method," *Int. J. Numer. Methods Eng.* **71**, 837–855 (2007).

[60] H. Cheng, L. Greengard, and V. Rokhlin, "A fast adaptive multipole algorithm in three dimensions," *J. Comput. Phys.* **155**, 468–498 (1999).

[61] L. Shen and Y. J. Liu, "An adaptive fast multipole boundary element method for three-dimensional potential problems," *Comput. Mech.* **39**, 681–691 (2007).

[62] Y. J. Liu and N. Nishimura, "The fast multipole boundary element method for potential problems: A tutorial," *Eng. Anal. Boundary Elements* **30**, 371–381 (2006).

[63] K. Yoshida, "Applications of fast multipole method to boundary integral equation method," Ph.D. dissertation, Department of Global Environment Engineering, Kyoto University (2001).

[64] W. H. Beyer, *CRC Standard Mathematical Tables and Formulae*, 29th ed. (CRC, Boca Raton, FL, 1991).

[65] L. Greengard and V. Rokhlin, "A new version of the fast multipole method for the Laplace equation in three dimensions," *Acta Numerica* **6**, 229–269 (1997).

[66] K. Yoshida, N. Nishimura, and S. Kobayashi, "Application of new fast multipole boundary integral equation method to crack problems in 3D," *Eng. Anal. Boundary Elements* **25**, 239–247 (2001).

[67] X. L. Chen and H. Zhang, "An integrated imaging and BEM for fast simulation of freeform objects," *Computer-Aided Design and Applications* **5**(1–4), 371–380 (2008).

[68] L. F. Greengard, M. C. Kropinski, and A. Mayo, "Integral equation methods for Stokes flow and isotropic elasticity in the plane," *J. Comput. Phys.* **125**, 403–414 (1996).

[69] L. F. Greengard and J. Helsing, "On the numerical evaluation of elastostatic fields in locally isotropic two-dimensional composites," *J. Mech. Phys. Solids* **46**, 1441–1462 (1998).

[70] J. D. Richardson, L. J. Gray, T. Kaplan, and J. A. Napier, "Regularized spectral multipole BEM for plane elasticity," *Eng. Anal. Boundary Elements* **25**, 297–311 (2001).

[71] T. Fukui, "Research on the boundary element method – Development and applications of fast and accurate computations," Ph.D. dissertation (in Japanese), Department of Global Environment Engineering, Kyoto University (1998).

[72] T. Fukui, T. Mochida, and K. Inoue, "Crack extension analysis in system of growing cracks by fast multipole boundary element method (in Japanese)," in *Proceedings of the Seventh BEM Technology Conference* (JASCOME, Tokyo, 1997), pp. 25–30.

[73] Y. J. Liu, "A new fast multipole boundary element method for solving large-scale two-dimensional elastostatic problems," *Int. J. Numer. Methods Eng.* **65**, 863–881 (2005).

[74] Y. J. Liu, "A fast multipole boundary element method for 2-D multi-domain elastostatic problems based on a dual BIE formulation," *Comput. Mech.* **42**, 761–773 (2008).

[75] P. Wang and Z. Yao, "Fast multipole DBEM analysis of fatigue crack growth," *Comput. Mech.* **38**, 223–233 (2006).

[76] Y. Yamada and K. Hayami, "A multipole boundary element method for two dimensional elastostatics," Report METR 95–07, Department of Mathematical Engineering and Information Physics, University of Tokyo (1995).

[77] V. Popov and H. Power, "An $O(N)$ Taylor series multipole boundary element method for three-dimensional elasticity problems," *Eng. Anal. Boundary Elements* **25**, 7–18 (2001).

[78] K. Yoshida, N. Nishimura, and S. Kobayashi, "Application of fast multi-pole Galerkin boundary integral equation method to crack problems in 3D," *Int. J. Numer. Methods Eng.* **50**, 525–547 (2001).

[79] Y.-S. Lai and G. J. Rodin, "Fast boundary element method for three-dimensional solids containing many cracks," *Eng. Anal. Boundary Elements* **27**, 845–852 (2003).

[80] Y. J. Liu, N. Nishimura, and Y. Otani, "Large-scale modeling of carbon-nanotube composites by the boundary element method based on a rigid-inclusion model," *Comput. Mater. Sci.* **34**, 173–187 (2005).

[81] Y. J. Liu, N. Nishimura, Y. Otani, T. Takahashi, X. L. Chen, and H. Munakata, "A fast boundary element method for the analysis of fiber-reinforced composites based on a rigid-inclusion model," *J. Appl. Mech.* **72**, 115–128 (2005).

[82] Y. J. Liu, N. Nishimura, D. Qian, N. Adachi, Y. Otani, and V. Mokashi, "A boundary element method for the analysis of CNT/polymer composites with a cohesive interface model based on molecular dynamics," *Eng. Anal. Boundary Elements* **32**, 299–308 (2008).

[83] V. Sladek and J. Sladek, eds., *Singular Integrals in Boundary Element Methods*, Advances in Boundary Element Series, C. A. Brebbia and M. H. Aliabadi, series eds. (Computational Mechanics Publications, Boston, 1998).

[84] S. Mukherjee, "Finite parts of singular and hypersingular integrals with irregular boundary source points," *Eng. Anal. Boundary Elements* **24**, 767–776 (2000).

[85] Y. J. Liu and F. J. Rizzo, "Scattering of elastic waves from thin shapes in three dimensions using the composite boundary integral equation formulation," *J. Acoust. Soc. Am.* **102**, 926–932 (1997).

[86] N. I. Muskhelishvili, *Some Basic Problems of Mathematical Theory of Elasticity* (Noordhoff, Groningen, The Netherlands, 1958).

[87] I. S. Sokolnikoff, *Mathematical Theory of Elasticity*, 2nd ed. (McGraw-Hill, New York, 1956).

[88] S. P. Timoshenko and J. N. Goodier, *Theory of Elasticity*, 3rd ed. (McGraw-Hill, New York, 1987).

[89] D. Gross and T. Seelig, *Fracture Mechanics with an Introduction to Micromechanics* (Springer, Dordrecht, The Netherlands, 2006).

[90] M. S. Ingber and T. D. Papathanasiou, "A parallel-supercomputing investigation of the stiffness of aligned, short-fiber-reinforced composites using the boundary element method," *Int. J. Numer. Methods Eng.* **40**, 3477–3491 (1997).

[91] I. G. Currie, *Fundamental Mechanics of Fluids* (McGraw-Hill, New York, 1974).

[92] C. Pozrikidis, *Boundary Integral and Singularity Methods for Linearized Viscous Flow* (Cambridge University Press, New York, 1992).

[93] H. Power and L. C. Wrobel, *Boundary Integral Methods in Fluid Mechanics* (Computational Mechanics Publications, Southampton, UK, 1995).

[94] J. Ding and W. Ye, "A fast integral approach for drag force calculation due to oscillatory slip stokes flows," *Int. J. Numer. Methods Eng.* **60**, 1535–1567 (2004).

[95] A. Frangi, "A fast multipole implementation of the qualocation mixed-velocity-traction approach for exterior Stokes flows," *Eng. Anal. Boundary Elements* **29**, 1039–1046 (2005).

[96] A. Frangi and A. D. Gioia, "Multipole BEM for the evaluation of damping forces on MEMS," *Comput. Mech.* **37**, 24–31 (2005).

[97] A. Frangi and J. Tausch, "A qualocation enhanced approach for Stokes flow problems with rigid-body boundary conditions," *Eng. Anal. Boundary Elements* **29**, 886–893 (2005).

[98] A. Frangi, G. Spinola, and B. Vigna, "On the evaluation of damping in MEMS in the slip–flow regime," *Int. J. Numer. Methods Eng.* **68**, 1031–1051 (2006).

[99] Y. J. Liu, "A new fast multipole boundary element method for solving 2-D Stokes flow problems based on a dual BIE formulation," *Eng. Anal. Boundary Elements* **32**, 139–151 (2008).

[100] A. Frangi and G. Novati, "Symmetric BE method in two-dimensional elasticity: Evaluation of double integrals for curved elements," *Comput. Mech.* **19**, 58–68 (1996).

[101] J. J. Perez-Gavilan and M. H. Aliabadi, "Symmetric Galerkin BEM for multi-connected bodies," *Commun. Numer. Methods Eng.* **17**, 761–770 (2001).

[102] L. Shen and Y. J. Liu, "An adaptive fast multipole boundary element method for three-dimensional acoustic wave problems based on the Burton–Miller formulation," *Comput. Mech.* **40**, 461–472 (2007).

[103] C. Pozrikidis, *Fluid Dynamics – Theory, Computation and Numerical Simulation* (Kluwer Academic, Boston, 2001).

[104] H. Power, "The interaction of a deformable bubble with a rigid wall at small Reynolds number: A general approach via integral equations," *Eng. Anal. Boundary Elements* **19**, 291–297 (1997).

[105] G. Zhu, A. A. Mammoli, and H. Power, "A 3-D indirect boundary element method for bounded creeping flow of drops," *Eng. Anal. Boundary Elements* **30**, 856–868 (2006).

[106] S. Mukherjee, S. Telukunta, and Y. X. Mukherjee, "BEM modeling of damping forces on MEMS with thin plates," *Eng. Anal. Boundary Elements* **29**, 1000–1007 (2005).

[107] H. A. Schenck, "Improved integral formulation for acoustic radiation problems," *J. Acoust. Soc. Am.* **44**, 41–58 (1968).

[108] A. J. Burton and G. F. Miller, "The application of integral equation methods to the numerical solution of some exterior boundary-value problems," *Proc. R. Soc. London Ser. A* **323**, 201–210 (1971).

[109] F. Ursell, "On the exterior problems of acoustics," *Proc. Cambridge Philos. Soc.* **74**, 117–125 (1973).

[110] R. E. Kleinman and G. F. Roach, "Boundary integral equations for the three-dimensional Helmholtz equation," *SIAM Rev.* **16**, 214–236 (1974).

[111] D. S. Jones, "Integral equations for the exterior acoustic problem," *Q. J. Mech. Appl. Math.* **27**, 129–142 (1974).

[112] W. L. Meyer, W. A. Bell, B. T. Zinn, and M. P. Stallybrass, "Boundary integral solutions of three-dimensional acoustic radiation problems," *J. Sound Vib.* **59**, 245–262 (1978).

[113] A. F. Seybert, B. Soenarko, F. J. Rizzo, and D. J. Shippy, "An advanced computational method for radiation and scattering of acoustic waves in three dimensions," *J. Acoust. Soc. Am.* **77**, 362–368 (1985).

[114] R. Kress, "Minimizing the condition number of boundary integral operators in acoustic and electromagnetic scattering," *Q. J. Mech. Appl. Math.* **38**, 323–341 (1985).

[115] A. F. Seybert and T. K. Rengarajan, "The use of CHIEF to obtain unique solutions for acoustic radiation using boundary integral equations," *J. Acoust. Soc. Am.* **81**, 1299–1306 (1987).

[116] K. A. Cunefare and G. Koopmann, "A boundary element method for acoustic radiation valid for all wavenumbers," *J. Acoust. Soc. Am.* **85**, 39–48 (1989).

[117] G. C. Everstine and F. M. Henderson, "Coupled finite element/boundary element approach for fluid structure interaction," *J. Acoust. Soc. Am.* **87**, 1938–1947 (1990).

[118] R. Martinez, "The thin-shape breakdown (TSB) of the Helmholtz integral equation," *J. Acoust. Soc. Am.* **90**, 2728–2738 (1991).

[119] K. A. Cunefare and G. H. Koopmann, "A boundary element approach to optimization of active noise control sources on three-dimensional structures," *J. Vib. Acoust.* **113**, 387–394 (1991).

[120] R. D. Ciskowski and C. A. Brebbia, *Boundary Element Methods in Acoustics* (Kluwer Academic, New York, 1991).

[121] G. Krishnasamy, T. J. Rudolphi, L. W. Schmerr, and F. J. Rizzo, "Hypersingular boundary integral equations: Some applications in acoustic and elastic wave scattering," *J. Appl. Mech.* **57**, 404–414 (1990).

[122] S. Amini, "On the choice of the coupling parameter in boundary integral formulations of the exterior acoustic problem," *Appl. Anal.* **35**, 75–92 (1990).

[123] T. W. Wu, A. F. Seybert, and G. C. Wan, "On the numerical implementation of a Cauchy principal value integral to insure a unique solution for acoustic radiation and scattering," *J. Acoust. Soc. Am.* **90**, 554–560 (1991).

[124] Y. J. Liu, "Development and applications of hypersingular boundary integral equations for 3-D acoustics and elastodynamics," Ph.D. dissertation, Department of Theoretical and Applied Mechanics, University of Illinois at Urbana-Champaign (1992).

[125] S.-A. Yang, "Acoustic scattering by a hard and soft body across a wide frequency range by the Helmholtz integral equation method," *J. Acoust. Soc. Am.* **102**, 2511–2520 (1997).

[126] V. Rokhlin, "Rapid solution of integral equations of scattering theory in two dimensions," *J. Comput. Phys.* **86**, 414–439 (1990).

[127] V. Rokhlin, "Diagonal forms of translation operators for the Helmholtz equation in three dimensions," *Appl. Comput. Harmon. Anal.* **1**, 82–93 (1993).

[128] M. Epton and B. Dembart, "Multipole translation theory for the three-dimensional Laplace and Helmholtz equations," *SIAM J. Sci. Comput.* **16**, 865–897 (1995).

[129] S. Koc and W. C. Chew, "Calculation of acoustical scattering from a cluster of scatterers," *J. Acoust. Soc. Am.* **103**, 721–734 (1998).

[130] L. Greengard, J. Huang, V. Rokhlin, and S. Wandzura, "Accelerating fast multipole methods for the Helmholtz equation at low frequencies," *IEEE Comput. Sci. Eng.* **5**(3), 32–38 (1998).

[131] M. A. Tournour and N. Atalla, "Efficient evaluation of the acoustic radiation using multipole expansion," *Int. J. Numer. Methods Eng.* **46**, 825–837 (1999).

[132] N. A. Gumerov and R. Duraiswami, "Recursions for the computation of multipole translation and rotation coefficients for the 3-D Helmholtz equation," *SIAM J. Sci. Comput.* **25**, 1344–1381 (2003).

[133] E. Darve and P. Havé, "Efficient fast multipole method for low-frequency scattering," *J. Comput. Phys.* **197**, 341–363 (2004).

[134] M. Fischer, U. Gauger, and L. Gaul, "A multipole Galerkin boundary element method for acoustics," *Eng. Anal. Boundary Elements* **28**, 155–162 (2004).

[135] J. T. Chen and K. H. Chen, "Applications of the dual integral formulation in conjunction with fast multipole method in large-scale problems

for 2D exterior acoustics," *Eng. Anal. Boundary Elements* **28**, 685–709 (2004).

[136] N. A. Gumerov and R. Duraiswami, *Fast Multipole Methods for the Helmholtz Equation in Three Dimensions* (Elsevier, Amsterdam, 2004).

[137] H. Cheng, W. Y. Crutchfield, Z. Gimbutas, L. F. Greengard, J. F. Ethridge, J. Huang, V. Rokhlin, N. Yarvin, and J. Zhao, "A wideband fast multipole method for the Helmholtz equation in three dimensions," *J. Comput. Phys.* **216**, 300–325 (2006).

[138] M. Abramowitz and I. A. Stegun, *Handbook of Mathematical Functions with Formulas, Graphs, and Mathematical Tables*, 10th ed. (United States Department of Commerce, U.S. Government Printing Office, Washington, DC, 1972).

[139] S. Marburg and T. W. Wu, "Treating the phenomenon of irregular frequencies," in S. Marburg and B. Nolte, eds., *Computational Acoustics of Noise Propagation in Fluids* (Springer, Berlin, 2008), pp. 411–434.

[140] Y. J. Liu and F. J. Rizzo, "Application of Overhauser $C^{(1)}$ continuous boundary elements to 'hypersingular' BIE for 3-D acoustic wave problems," in C. A. Brebbia and G. S. Gipson, eds., *Boundary Elements XIII* (Computation Mechanics Publications, Tulsa, OK, 1991), pp. 957–966.

[141] A. Messiah, "Clebsch–Gordan (C-G) Coefficients and '3j Symbols,'" in *Quantum Mechanics, Appendix C.I.* (North-Holland Amsterdam, The Netherlands, 1962), pp. 1054–1060.

[142] M. Bapat, L. Shen, and Y. J. Liu, "An adaptive fast multipole boundary element method for 3-D half-space acoustic wave problems," *Eng. Anal. Boundary Elements*, in press (2009).

[143] S. H. Chen and Y. J. Liu, "A unified boundary element method for the analysis of sound and shell-like structure interactions. I. Formulation and verification," *J. Acoust. Soc. Am.* **103**, 1247–1254 (1999).

[144] S. H. Chen, Y. J. Liu, and X. Y. Dou, "A unified boundary element method for the analysis of sound and shell-like structure interactions. II. Efficient solution techniques," *J. Acoust. Soc. Am.* **108**, 2738–2745 (2000).

[145] T. W. Wu, ed., *Boundary Element Acoustics: Fundamentals and Computer Codes* (WIT Press, Southampton, UK, 2000).

Index

Printed in the United States
By Bookmasters